Signals and Communication Technology

This series is devoted to fundamentals and applications of modern methods of signal processing and cutting-edge communication technologies. The main topics are information and signal theory, acoustical signal processing, image processing and multimedia systems, mobile and wireless communications, and computer and communication networks. Volumes in the series address researchers in academia and industrial R&D departments. The series is application-oriented. The level of presentation of each individual volume, however, depends on the subject and can range from practical to scientific.

Indexing: All books in "Signals and Communication Technology" are indexed by Scopus and zbMATH

For general information about this book series, comments or suggestions, please contact Mary James at mary.james@springer.com or Ramesh Nath Premnath at ramesh.premnath@springer.com.

Amit Kumar Mishra · Vimal Bhatia
Editors

Integrated Sensing and Communication for Sustainable Development Goals

Springer

Editors
Amit Kumar Mishra
Faculty of Space Technologies
AGH University of Kraków
Kraków, Poland

Vimal Bhatia
Department of Electrical Engineering
Indian Institute of Technology Indore
Indore, Madhya Pradesh, India

ISSN 1860-4862 ISSN 1860-4870 (electronic)
Signals and Communication Technology
ISBN 978-981-92-0208-9 ISBN 978-981-92-0209-6 (eBook)
https://doi.org/10.1007/978-981-92-0209-6

This Springer imprint is published by the registered company Springer Nature Singapore Pte Ltd.
The registered company address is: 152 Beach Road, #21-01/04 Gateway East, Singapore 189721, Singapore

Preface

Radar and radio frequency (RF) telecommunication systems are similar yet dissimilar in many aspects and applications. Since the early 1960s, there have been initiatives to merge the functionality of radio sensing and radio telecommunications using shared resources. However, the research in this area picked up speed with 3GPP standards adopting shared RF sensing and communication in standards as an important pillar for 6G and beyond. The domain, currently known as integrated sensing and communication (ISAC) or also as joint sensing and communications (JSAC), is witnessing aggressive research and development initiatives throughout the world with both academia and industry keen on adopting the joint functionality. This monograph endeavours to put together some of the emerging directions in ISAC and some interesting applications. We expect the book to be of use to both industry engineers as well academic researchers.

Kraków, Poland
Indore, India

Prof. Amit Kumar Mishra
Prof. Vimal Bhatia

Contents

Chapter 1
5G-Based Integrated Communication and Sensing for Situational Awareness

Tomasz P. Zieliński, Karol Abratkiewicz, Radosław Maksymiuk, Piotr Samczyński, Jacek Wszołek, Marek Wypich, and Christof Schüpbach

Abstract The fifth-generation (5G) communication has recently gained massive attention due to its flexibility and efficacy, allowing high-throughput data streaming. 5G networks are also considered promising candidates for future passive radar illuminators. Integrating 5G networks with active and passive radar systems featuring communication functions represents a transformative approach in modern sensing and data transmission technologies. The unique properties of 5G, including high bandwidth, low latency, and dense network deployments, make it well suited for integrated communication and sensing systems. When joined, the systems benefit from improved detection accuracy, real-time data transmission, and more efficient spectrum utilization. However, challenges remain, such as content-dependent transmission, beamforming, transceiver radiation pattern, and irregular network organization, to name only a few. This chapter explores the benefits, technical considerations,

---

T. P. Zieliński · J. Wszołek · M. Wypich
AGH University of Krakow, Institute of Telecommunications and Cybersecurity, Krakow, Poland
e-mail: tomasz.zielinski@agh.edu.pl

J. Wszołek
e-mail: jacek.wszolek@agh.edu.pl

M. Wypich
e-mail: marekwypich@student.agh.edu.pl

K. Abratkiewicz (✉) · R. Maksymiuk · P. Samczyński
Institute of Electronic Systems, Warsaw University of Technology, Warsaw, Poland
e-mail: karol.abratkiewicz@pw.edu.pl

R. Maksymiuk
e-mail: marekwypich@agh.edu.pl

P. Samczyński
e-mail: piotr.samczynski@pw.edu.pl

J. Wszołek · M. Wypich
Ericsson POL, Krakow, Poland

C. Schüpbach
Armasuisse Science and Technology, Thun, Switzerland
e-mail: christof.schuepbach@armasuisse.ch

A. K. Mishra and V. Bhatia (eds.), *Integrated Sensing and Communication for Sustainable Development Goals*, Signals and Communication Technology,
https://doi.org/10.1007/978-981-92-0209-6_1

and future opportunities for leveraging 5G networks in joint radar and communication applications.

Keywords 5G · 5G-based communication and sensing · 5G passive radar · Radar signal processing · Active and passive radar

List of Abbreviations

3GPP	3rd Generation Partnership Project
5G	Fifth Generation (mobile network/standard)
5G NR	5G New Radio
A1, A2	Transmit antennas 1 and 2
BCH	Broadcast Channel
BPSK	Binary Phase Shift Keying
BTS	Base Transceiver Station
CAF	Cross-Ambiguity Function
CFO	Carrier Frequency Offset
CIR	Channel Impulse Response
CFR	Channel Frequency Response
CP	Cyclic Prefix
CPU	Central Processing Unit
COTS	Commercial Off-The-Shelf
CSI	Channel State Information
CSI-RS	Channel State Information Reference Signal
DCI	Downlink Control Information
DDM	Delay-Doppler Map
DFT	Discrete Fourier Transform
DM-RS	Demodulation Reference Signal
ECA-CD	Extensive Cancellation Algorithm by Carrier and Doppler
FDD	Frequency-Division Duplex/Frequency-Division Duplexing
FEQ	Frequency Equalization
FFT	Fast Fourier Transform
FDM	Frequency-Division Multiplexing
GNSS	Global Navigation Satellite System
GSCN	Global Synchronization Channel Number
gNB	Next-Generation Node B/5G Base Station
HARQ	Hybrid Automatic Repeat reQuest
IDFT	Inverse Discrete Fourier Transform
IFFT	Inverse Fast Fourier Transform
ISAC	Integrated Sensing and Communication
L1	Layer-1 (Physical Layer)
L2	Layer-2 (MAC/RLC/PDCP Layer)
L3	Layer-3 (RRC/SDAP Layer)

LDPC	Low-Density Parity-Check
MAC	Medium Access Control
MIB	Master Information Block
MIMO	Multiple-Input Multiple-Output
MLS	Maximum Length Sequence
mmWave	Millimeter Wave
NR	New Radio (5G standard)
NSA	Non-Standalone (5G architecture)
OFDM	Orthogonal Frequency-Division Multiplexing
PAPR	Peak-to-Average Power Ratio
PBCH	Physical Broadcast Channel
PC	Personal Computer
PDCCH	Physical Downlink Control Channel
PDCP	Packet Data Convergence Protocol
PDSCH	Physical Downlink Shared Channel
PDU	Protocol Data Unit
PHY	Physical Layer
PNFR	Peak-to-Noise Floor Ratio
PRACH	Physical Random Access Channel
PRS	Positioning Reference Signal
PRB	Physical Resource Block
PSS	Primary Synchronization Signal
PUCCH	Physical Uplink Control Channel
PUSCH	Physical Uplink Shared Channel
PT-RS	Phase Tracking Reference Signal
QAM	Quadrature Amplitude Modulation
QoS	Quality of Service
QPSK	Quadrature Phase Shift Keying
RAM	Random Access Memory
RB	Resource Block
RE	Resource Element
RF	Radio Frequency
RLC	Radio Link Control
RRC	Radio Resource Control
RoD	Radar-on-Demand
RV	Range-Velocity
RVM	Radar-Velocity Map
RX	Receiver
SCS	Sub-Carrier Spacing
SDAP	Service Data Adaptation Protocol
SDG	Sustainable Development Goal
SDR	Software-Defined Radio
SNR	Signal-to-Noise Ratio
SR	Scheduling Request
SRS	Sounding Reference Signal

SSB	Synchronization Signal Block
SSS	Secondary Synchronization Signal
STFT	Short-Time Fourier Transform
TDD	Time Division Duplex
TV-CFR	Time-Variant Channel Frequency Response
TV-CIR	Time-Variant Channel Impulse Response
TX	Transmitter
UE	User Equipment
USRP	Universal Software Radio Peripheral

1.1 Introduction

Fifth-generation (5G) cellular networks have recently garnered significant interest thanks to their high-speed data transmission and low latency capabilities. 5G has developed advanced features such as massive multiple-input multiple-output (MIMO) antenna arrays, wide bandwidths, millimeter-wave (mmWave) frequency utilization, and modern transmission techniques to meet the increasing demands for efficient data exchange.

As with many wireless and broadcasting technologies, 5G has become an appealing subject in both passive and active radar researches. Numerous recent studies confirm its potential, demonstrating that 5G can be effectively integrated into radar systems for improved performance and functionality [1–6]. Typically, 5G base stations are considered either cooperative or non-cooperative illumination sources [1, 7], or 5G signals are emulated using software-defined radio (SDR) in controlled environments [8, 9]. Owing to its wide transmission bandwidth, the 5G standard is particularly well suited for sensing applications alongside communications, forming the foundation for integrated sensing and communication (ISAC) systems [10, 11].

ISAC aims to merge data transmission and environmental sensing within a unified architecture, utilizing the same radio signals to perform both tasks. This dual use reduces spectrum occupation and hardware complexity. By applying advanced waveform design and signal processing, ISAC systems can enhance detection, tracking, and localization while preserving high data rates. Such integration is highly advantageous in areas like autonomous navigation, urban infrastructure, and traffic control. However, achieving an optimal balance between communication performance and sensing accuracy poses significant design challenges, particularly in terms of resource allocation and interference management.

From a broader perspective, integrating communication and sensing within the same physical infrastructure is particularly promising in civilian applications. This approach aligns closely with global initiatives toward achieving the United Nations Sustainable Development Goals (SDGs), especially in areas where reliable data acquisition and connectivity play a critical role. By enabling efficient environmental monitoring, real-time data collection, and low-power communication, integrated systems such as ISAC can support efficient urban infrastructure protection and general situational awareness, among others. These capabilities are particularly impactful in resource-constrained environments where infrastructure must be both cost-effective and energy-efficient.

In line with the goals of this book, which examines how integrated systems can enhance monitoring, data fusion, wireless protocols, and energy-aware design, this chapter focuses on the development of energy-efficient dual-function systems that operate in a shared frequency band. In particular, we explore how such systems can simultaneously detect environmental features (e.g., obstacles, terrain variation) and exchange information using the same waveform and hardware platform. This approach reduces power consumption and the required hardware infrastructure while maintaining high performance in communication and sensing tasks. The resulting solutions contribute to enhanced situational awareness and sustainable deployment in applications such as smart agriculture, environmental monitoring, and early warning systems.

This chapter has the following structure: Sect. 1.2 describes the principles of the 5G standard from the perspective of its utility on ISAC systems. Next, in Sect. 1.3, the concept of the 5G-based ISAC system is presented. The simulations and real-life signal analysis results are delineated in Sect. 1.4. The chapter closes with a summary and prospects in Sect. 1.5.

1.2 The Why and How of 5G Signal

1.2.1 OFDM-Based Digital Transmission Basics

The orthogonal frequency-division multiplexing (OFDM) is used in the 5G new radio (NR) cellular network to access the radio frequency (RF) transmission medium [12]. Digital bit communication exploiting OFDM is explained in Fig. 1.1.

In the **OFDM transmitter**, a bitstream "0101111000" is cut into pieces. If our single sub-carrier will "carry" only $N_{\text{bits}} = 2$ bits, cutting will give: 01 | 01 | 11 | 10 | 00. Then bit cuts are interpreted as only four possible binary numbers [00, 01, 10, 11], which equals to $0, 1, 2, 3$ in decimal numbers (in general: $0, 1, 2, ..., 2^{N_{\text{bits}}} - 1$). In our example, we obtain 1,1,3,2,0. Now, decimal numbers are interpreted as indexes in the modulation table in which some complex numbers are stored, for example, for quadrature amplitude modulation (QAM) with four carrier states, i.e., 4-QAM $= [1 + j, 1 - j, -1 + j, -1 - j]$ or equivalently, neglecting $\sqrt{2}$, $[e^{j\frac{\pi}{4}}, e^{-j\frac{\pi}{4}}, e^{j\frac{3\pi}{4}}, e^{-j\frac{3\pi}{4}}]$. Therefore, different table indexes are used depending on bit values, and different complex numbers are taken from the QAM table. In our example: $[e^{-j\frac{\pi}{4}}, e^{-j\frac{\pi}{4}}, e^{-j\frac{3\pi}{4}}, e^{j\frac{3\pi}{4}}, e^{j\frac{\pi}{4}}]$. In Fig. 1.1, these values are denoted as x_k and put into columns of the matrix $\mathbf{X}$. The columns, having N_{FFT} elements (e.g., 3 elements in Fig. 1.1), are called the OFDM symbols. All OFDM symbols of the matrix give together one transmission frame, for example, of 4G/5G communication services. Then inverse discrete Fourier transform (DFT) is performed

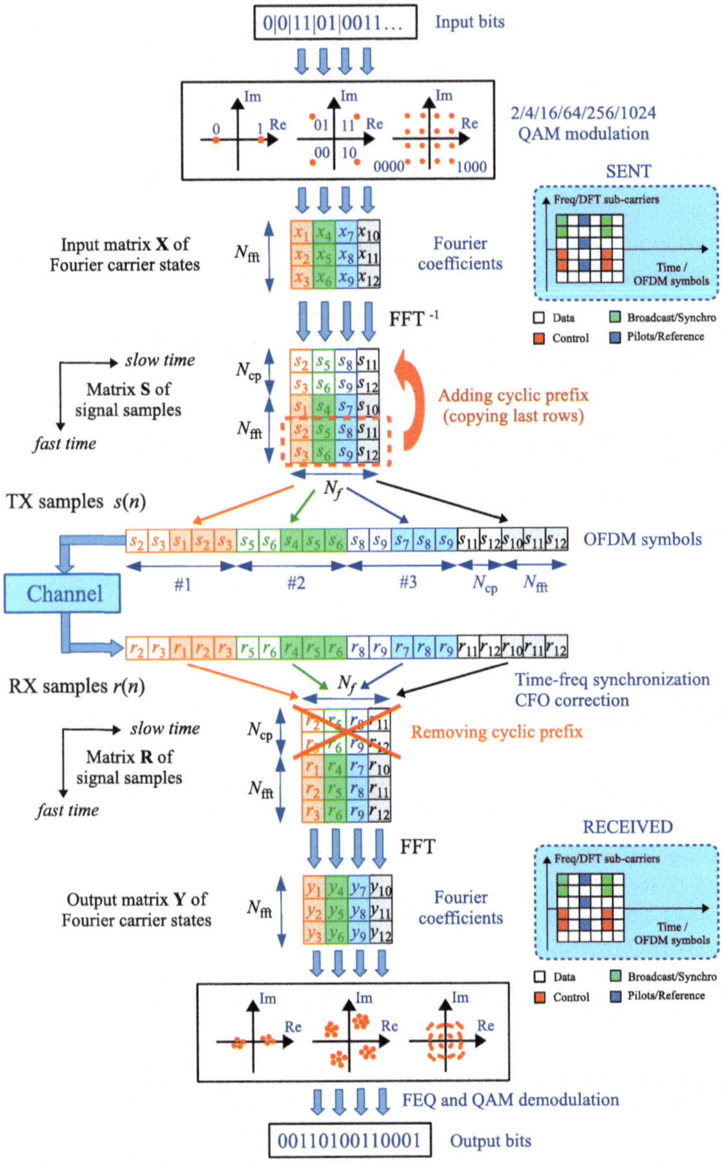

Fig. 1.1 OFDM-based digital communication fundamentals: information is transmitted in Fourier coefficients of harmonic carriers combined into OFDM symbols (consecutive Fourier spectra)

over columns and spectral Fourier coefficients are transformed into signal samples s_k, columns of the matrix **S**: one signal being summation of N_{FFT} Fourier harmonics $f_k(n) = e^{j\frac{2\pi}{N}kn}$, $n = 0, 1, ..., N_{\text{FFT}} - 1$, where $k = 0, 1, ..., N_{\text{FFT}} - 1$ denotes the harmonic index and decides about the frequency of harmonic oscillation. These harmonics are scaled by coefficients taken from the QAM table, and they *carry* information about their indexes and bits associated with them. In order to eliminate OFDM inter-symbol interference, caused during transmission by a fading channel, the N_{CP} last signal samples of each column are copied to the beginning of the column. They are called a cyclic prefix (CP). The CP length N_{CP} should be longer than the channel impulse response. Finally, all samples from the **S** matrix are *serialized*, i.e., put one-by-one column-by-column, and transmitted.

In the **OFDM receiver** all operations are reversed: after frame and symbol synchronization as well as carrier frequency offset (CFO) compensation, received signal samples r_k are put into columns of the matrix **R** ($N_{\text{CP}} + N_{\text{FFT}}$ elements in each column). Then, samples belonging to the cyclic prefix are removed, DFTs are performed on matrix columns, and the matrix **Y** is obtained with the Fourier coefficients y_k. However, due to channel disturbance, the received numbers y_k are not the same as the transmitted numbers x_k. Therefore, channel influence has to be identified and equalized (at least minimized). After channel estimation and frequency equalization (FEQ), the estimated transmitted carrier states $\hat{x}_k$ are found, e.g., values from the applied modulation 4-QAM table $[1 + j, 1 - j, -1 + j, -1 - j]$. Finally, the sequence of indexes of the carrier states is changed to a bit sequence, and "0101111000" should be obtained.

Channel estimation and equalization in OFDM systems has a crucial role. Let us introduce the following denotations referring to Fig. 1.1: $X_l(k)$, the k-th element in the l-th column of matrix **X**, is the DFT coefficient of k-th carrier in the l-th DFT spectrum (similarly $Y_l(k)$ for matrix **Y**), and $\mathbf{k}_{all}, \mathbf{k}_d, \mathbf{k}_p$ are indexes of: all used carriers, carriers with transmitted data and control information, and carriers with transmitted reference (pilot) values, respectively. The pilot carriers have precisely specified amplitude-phase states $X_l(\mathbf{k}_p)$, which are periodically sent and used for system synchronization, channel estimation, and equalization. The receiver, aware of both the received and transmitted signal spectra, i.e., $Y_l(\mathbf{k}_p)$ and $X_l(\mathbf{k}_p)$, respectively, divides them to determine the channel frequency response (CFR) for the pilot harmonics, i.e., $H_l(\mathbf{k}_p) = Y_l(\mathbf{k}_p)./X_l(\mathbf{k}_p)$. The CFR is then interpolated for all relevant frequencies, including those used for data transmission ($H_l(\mathbf{k}_p) \rightarrow H_l(\mathbf{k}_{all})$). This interpolated CFR is utilized to equalize the data transmitted via data carriers by dividing the spectrum of the received signal by the CFR: $\hat{X}_l(\mathbf{k}_d) = Y_l(\mathbf{k}_d)./H_l(\mathbf{k}_d)$.

In summary, the OFDM used in 5G NR is a multi-carrier transmission technique in which bits are coded in discrete Fourier transform (DFT) coefficients and a multi-harmonic signal, resulting from the inverse DFT, is transmitted through a communication channel. The signal processing path from transmitter to receiver consists of the following steps:
Bits $\rightarrow$ Fourier coefficients $\rightarrow$ inverse DFT $\rightarrow$ addition of the CP to the signal $\rightarrow$ signal transmission through a wireless channel $\rightarrow$ time and frequency synchro-

nization (including CFO correction) → CP removal → DFT → channel estimation using reference signals (pilots) → correction of DFT coefficients of sent data → bit detection.

1.2.2 5G NR Main Components

5G hardware components and their connection are presented in Fig. 1.2. The 5G network consists of a radio part, which includes user equipment (UE) and base stations (gNB), and a core part, commonly called the 5G Core. Transmission gNB to UE is called a downlink, UE-to-gNB—an uplink, and UE-to-UE—a sidelink. The interface between the gNB and the 5G Core is technically referred to as the backhaul link.

5G hierarchical software organization is shown in Fig. 1.3. The 5G NR system is organized into three **layers** (L1, L2, L3) which have different specialization (bottom-up): (1) Layer-1 bottom: PHY (PHYsical) aspects, (2) Layer-2 middle: MAC (Medium Access Control), RLC (Radio Link Control), and PDCP (Packet Data Convergence Protocol). The PDCP layer delivers basic connectivity to the higher layers. For the control-plane protocol stack, it provides connectivity to the top (3) Layer-3: RRC (Radio Resource Control) protocol. However, for the user-plane protocol stack, connectivity is provided to the newly introduced 5G SDAP (Service Data Adaptation Protocol) layer, which is responsible for the 5G QoS (Quality of Service) delivery. Data transfer between layers is performed using **channels: logical, transport, and physical**. Logical channel delivers RLC PDUs (Protocol Data Units) to the MAC layers. After receiving RLC PDUs, which originate from many different sources, the MAC layer decides which are supposed to be scheduled and when. Scheduling decisions are realized by formulating MAC PDUs, which are then transferred to the physical layer through the transport channel. The last, Physical Channels, carry information over the air interface.

There are three main types of physical **channels: broadcast, control, and data ones**. The broadcast channel allows UEs access to the 5G network, while the control channel is exploited to send system information concerning allocating available resources. In the middle MAC+PDCP layer, channel data are converted from the physical format of the PHY layer to the logical format of the RRC layer (in bottom-up direction) and vice versa (in top-down direction).

Control channels use specified resource blocks (RB) for strictly specified exchange (via hand-shaking) of control information and data between 5G providers and 5G users (1 RB = information carried by IQ states of 12 consecutive subcarriers of one OFDM symbol, i.e., by 12 DFT coefficients). The following channels are defined: (1) downlink (gNB→UE) and uplink (UE→gNB) (sidelink UE→UE is now out of our interest), (2) control and data, (3) broadcast (in downlink the basic system information is sent which is necessary for an UE to request accessing the network) and random access (in uplink, the UE requests the mentioned above access).

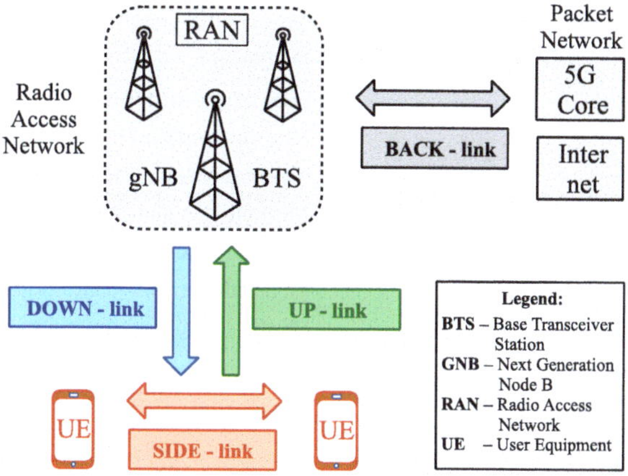

Fig. 1.2 5G hardware components and their co-operation: radio (gNB/BTS—base station) and core network connected via backhaul, UE—user equipment connected to gNB with the down/up/side/back-links

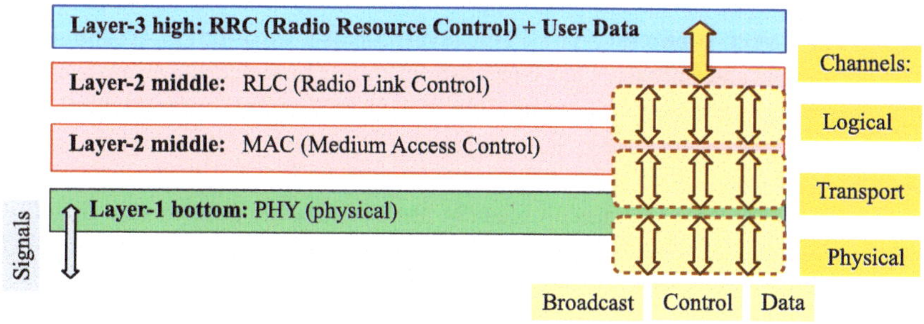

Fig. 1.3 5G software organization: layers, channels, and signals

Physical signals are strictly defined by in-phase (I) and quadrature (Q) values (jointly (IQ) values), generated by pseudo-random generators, lying in strictly defined time–frequency 5G grid positions. They are used to synchronize and transmit the most general system information, like the radio cell identification number. Both values and their localization are known to the receivers. Since the receiver knows what has been sent and what is received, it can perform time–frequency synchronization and estimate different channel disturbances (e.g., attenuation and phase shift) and hardware distortions (e.g., CFO and phase noise). Typically, CFR is estimated at some points of the time–frequency grid using physical signals and then interpolated for all carriers and OFDM symbols. In the further part of this chapter, only the 5G NR physical layer is of interest to us.

Table 1.1 Documentation of 5G NR physical layer

3GPP TS number	Description
38.201	General information [13]
38.202	Services provided by physical layer [14]
38.211	Physical channels and modulation [15]
38.212	Multiplexing and channel coding [16]
38.213	Physical layer procedures for control [17]
38.214	Physical layer procedures for data [18]
38.215	Physical layer measurements [19]

1.2.3 5G NR Physical Layer

Documentation. The physical layer of 5G NR is specified in the 3GPP TS 38.2xx documents listed in Table 1.1 [12]. It uses the channels and reference signals listed in Fig. 1.4. Their acronyms and main applications are given in this figure as well.

Physical Layer Elements. The PHY layer consists of different downlink and uplink channels and signals. One can observe that: (1) synchronization and network access signals use low-level modulations (BPSK—Binary Phase Shift Keying, QPSK—Quadrature Phase Shift Keying), (2) synchronization signals are not channel-coded (there is no sense doing this—we are only interested in analog values of carrier states), (3) control channels are coded with polar codes (in 4G—convolutional codes) while shared channels use low-density parity-check (LDPC) (in 4G—turbo codes).

It is important to note that the signals sent between gNB and UE, in both directions, play a crucial role in the PHY layer. If they are used for synchronization with the 5G grid, such as PSS and SSS signals, their positions on this grid are unknown. They should be found using cross-correlation of the received waveform with known samples of reference signals (in general, correlation can be performed in both the time and frequency domains). Different channel disturbances and hardware distortion can be identified and corrected by knowing the sent reference values of different, appropriately chosen pilot signals and the received deformed ones.

Physical Downlink Channels

1. **Physical Downlink Shared Channel (PDSCH)** is a high-speed channel, which is used for transmission of user data, exploiting QPSK, 16QAM, 64QAM, and 256QAM modulations.
2. **Physical Downlink Control Channel (PDCCH)** is a high reliability channel, carrying the Downlink Control Information (DCI) which includes scheduling schemes used for transmission of downlink (PDSCH) and uplink (PUSCH) data. It makes use of the QPSK modulation.

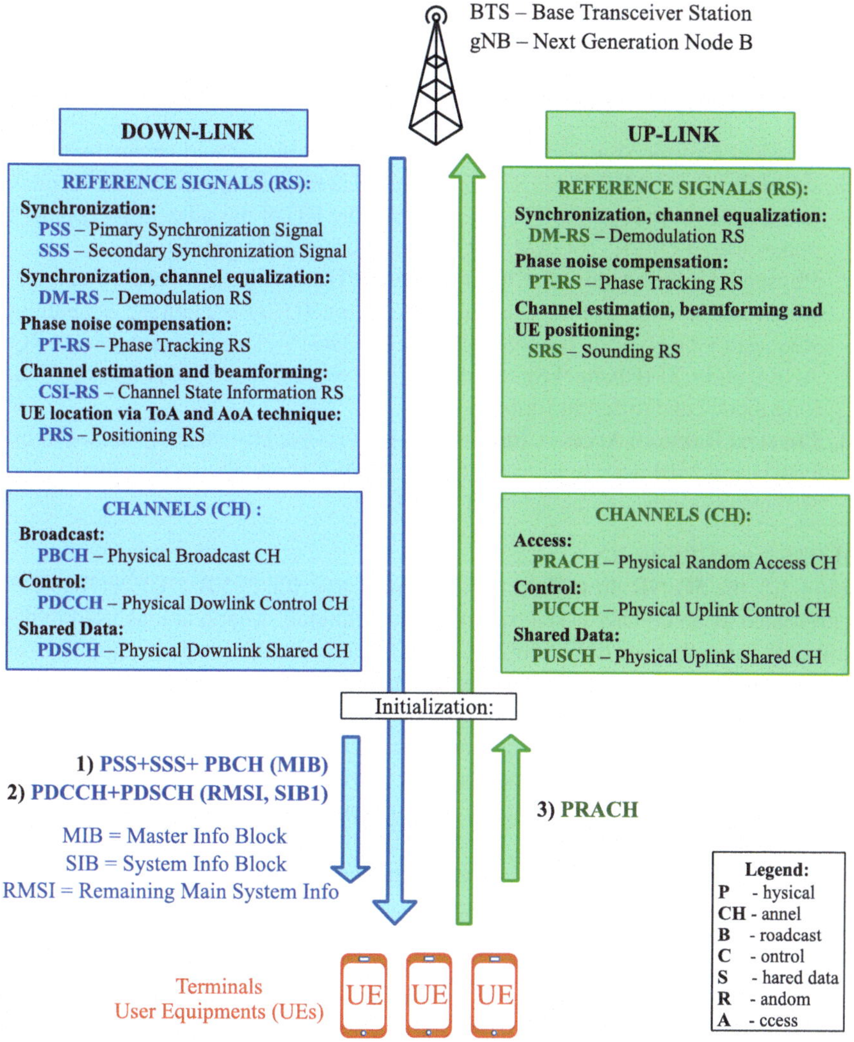

Fig. 1.4 5G physical layer: signals and channels used in downlink and uplink. Signals and control channels are the best candidates for the passive radar illuminator

3. **Physical Broadcast Channel (PBCH)** is the most reliable and the only always-on channel in the 5G NR system, which is used for RF cell search, cell synchronization, and beam sweeping procedures. It is part of the Synchronization Signal Block (SSB), consisting of two synchronization signals (primary and secondary synchronization signals—PSS and SSS, respectively) and the PBCH. The PBCH carries the Master Information Block (MIB) containing the basic system infor-

mation necessary for the UE to access the 5G network. PBCH makes use of the QPSK modulation.

Physical Uplink Channels

1. **Physical Uplink Shared Channel (PUSCH)** is a high-speed channel which is scheduled by the gNB MAC Scheduler and used for transmission of the user data to the base station (gNB). The DCI informs the UE of the scheduling results. It makes use of QPSK, 16QAM, 64QAM, and 256QAM modulations.
2. **Physical Uplink Control Channel (PUCCH)** is a high reliability channel, which carries the Uplink Control Information (UCI) containing Scheduling Request (SR), Hybrid Automatic Repeat ReQuest Acknowledgement (HARQ-ACK), and CSI (Channel State Information) reports. The PUCH has short and long modes and different formats.
3. **Physical Random Access Channel (PRACH)** is used by the UE (after receiving PBCH and MIB decoding) to request from gNB access to the 5G network. It is also utilized in beam management procedures.

OFDM-Based TX-to-RX

In Fig. 1.5, the **5G NR downlink** (base station to user link, gNB-2-UE, gNB→UE) is presented as an OFDM-based transmission technique. Abbreviations used in this figure have already been defined in Fig. 1.4. It is important to note that a signal passed in 5G by a telecommunication channel results from the computation of a sequence of inverse DFTs (IDFTs), namely inverse fast Fourier transforms (IFFTs) in practical implementation, and has a time–frequency (OFDM symbol—DFT sub-carrier) structure. Each element of the time–frequency grid is called in 5G a *Resource Element* (RE), and 12 such consecutive vertical elements (for 12 sub-carriers in one OFDM symbol) give a Physical Resource Block (PRB). The user plane Data is always allocated at the PRB level, while signaling (Control Channels and Signals) can be assigned at the RE level. Among others, the Downlink Control Information (DCI) specifies PRBs that are allocated to the data of each 5G user. Since the receiver should calculate IDFTs on signal samples belonging to many separate OFDM symbols, it should first synchronize with signal samples of these blocks (OFDM *symbol synchronization*). For this purpose, the CP can be used. Since OFDM symbols are collected (grouped) into blocks called frames, the *frame synchronization* should also be performed. For this, the SSB is used, consisting of the PSS, SSS, and PBCH (containing information that is necessary for accessing the 5G network by its users). In addition to PSS and SSS, several other physical signals are incorporated into the 5G time–frequency grid. Their names and applications were already presented.

The **uplink 5G processing scheme** (user to base station, UE-2-gNB, UE→gNB) is very similar to the downlink scheme, presented in Fig. 1.5. Of course, uplink channels and signals are used in this case, not the downlink ones. Simplifying, the main difference is that an additional DFT (FFT) can be performed on the IQ symbols before the resource mapping in the transmitter, and an additional IDFT (IFFT) can be calculated after the resource mapping in the receiver. This technique, called a

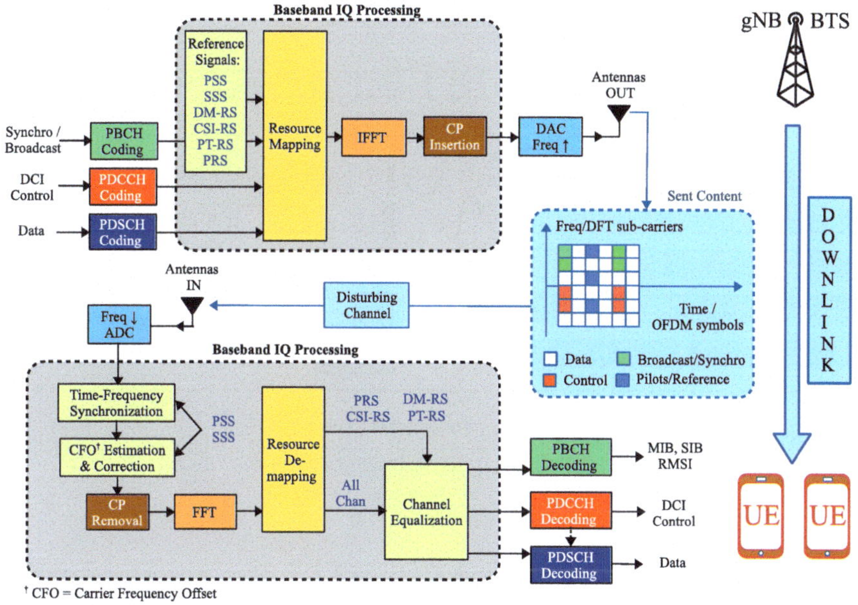

Fig. 1.5 The 5G (NR) downlink (gNB to UE) transmission scheme as an OFDM-based digital communication

DFT-spread-OFDM, aims to reduce the peak-to-average power ratio (PAPR).

5G Frame

The 5G frame structure is explained in Fig. 1.6. The frame lasts 10 ms and consists of 10 sub-frames, 1 ms each. According to applied *numerology*, defined by the parameter μ, each sub-frame is partitioned (divided) into different numbers of slots (1, 2, 8, 16, 64, 256). In turn, each slot is built from 14 OFDM symbols—the first one has a more extended CP than the rest.

When the OFDM symbol in the slot becomes shorter, the period of its fundamental harmonic also becomes shorter, and the fundamental frequency $f_0 = \Delta f$ of the symbol DFT becomes higher, i.e., the sub-carrier spacing (SCS) is increasing (see central part of Fig. 1.6). As a consequence, the atomic element of the 5G time–frequency grid is changing its time–frequency shape/support: it becomes narrower in time direction and wider in frequency direction (i.e., the OFDM symbols are repeated more often but they *consume* a wider frequency band—see Fig. 1.7).

Finally, in Table 1.2, a specification of the physical 5G frame is given. We see that the 5G standard also provides support for the 4G setup: $\Delta f = 15$ kHz, sampling rate 30.72 MHz, time slot duration 1 millisecond, and its division into 14 OFDM symbols, OFDM symbol duration 66.67 μs, and cyclic prefix length equal to 4.69 μs. The increase in the spacing of the sub-carrier 2^n times causes the same increase in

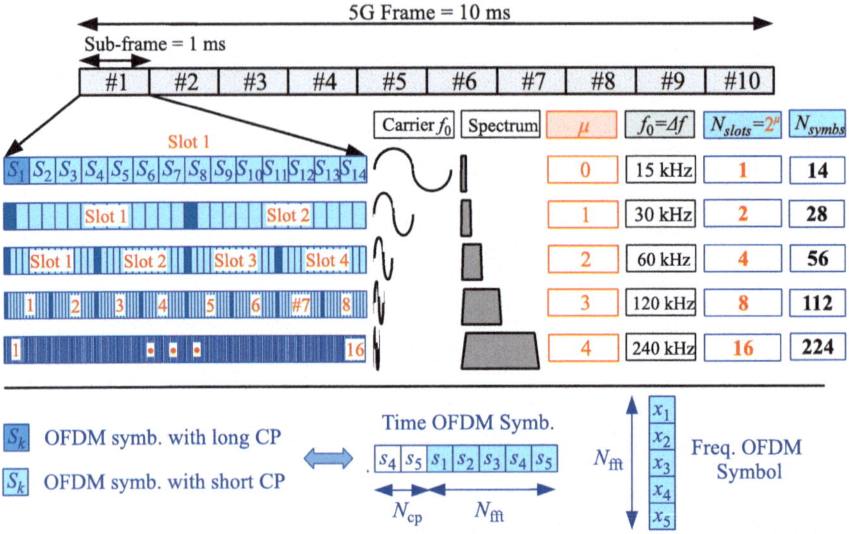

Fig. 1.6 5G frame: consisting of 10 sub-frames, 1 ms each. Depending on the numerology, each sub-frame consists of one or more slots. Each slot has 14 OFDM symbols. Higher numerology is usually used for wider channels. Default C-band numerology is $\mu = 1$

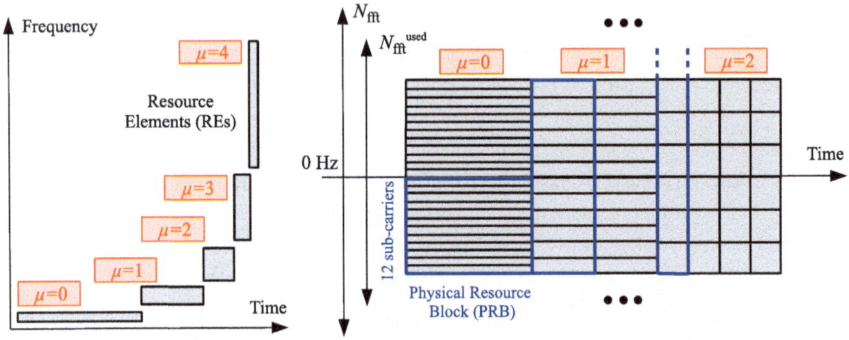

Fig. 1.7 Numerology impact on NR symbol bandwidth and duration: the higher the numerology (μ), the wider sub-carrier spacing (symbol bandwidth) and the shorter symbol duration

the slots in the sub-frame. It is clearly visible in Fig. 1.6, where 5G frames are shown for different values Δf (resulting from the sampling rate used).

The 5G NR downlink and uplink transmission can be organized in two ways: as a **Frequency-Division Duplex (FDD)** and a **Time Division Duplex (TDD)**—see Fig. 1.8. In the FDD technique, separate frequency bands are used for two directions. In TDD time slots (14 OFDM symbols) are combined into blocks of slots, and each OFDM symbol in the block can be defined as: a downlink one (D), an uplink one (U), and a switching one (S)—therefore, many possible formats are available. Switch-

Table 1.2 5G New Radio physical downlink and uplink specification

Sub-carrier spacing Δf [kHz]	15	30	60	120
OFDM symbol duration [μs]	66.67	33.33	16.67	8.33
Cyclic prefix duration [μs]	4.69	2.34	1.17	0.586
Slot duration (14 symbols) [μs]	1000	500	250	125
Slots in one sub-frame	1	2	4	4
Sampling rate [MHz]	30.72	61.44	122.88	245.76

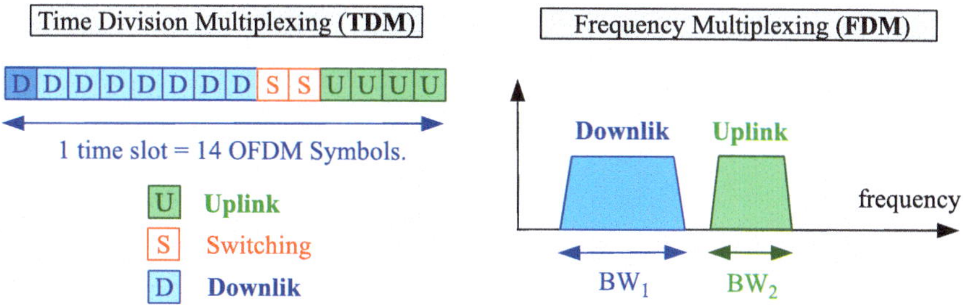

Fig. 1.8 Multiplexing of downlink and uplink channels in 5G: time division multiplexing (TDM) (left) or frequency-division multiplexing (FDM) (right)

ing slots are used for time separation (as time guard zones) between downlink and uplink transmission periods. All devices, gNB and UEs, should be synchronized—for this purpose, the Global Navigation Satellite System (GNSS) systems and/or other synchronization methods like packet layer Precision Time Protocol (IEEE 1588v2), physical SyncE (Synchronous Ethernet), or Sync uW (Sync over Microwave) are used.

5G Signal Synchronization Block

In 5G, the base station generates the SSB, presented in Fig. 1.9.

The SSB consists of:

1. The primary PSS synchronization signal, being a modulated maximum length sequence (MLS) having 127 samples (not Zadoff–Chu sequence as in 4G LTE); there are three different PSS sequences defined ($N_{ID}^{(2)} = 0, 1, 2$).
2. The secondary SSS synchronization signal, being also a MLS sequence having 127 samples; there are 336 different SSS sequences defined ($N_{ID}^{(1)} = 0, 1, 2, ..., 365$).

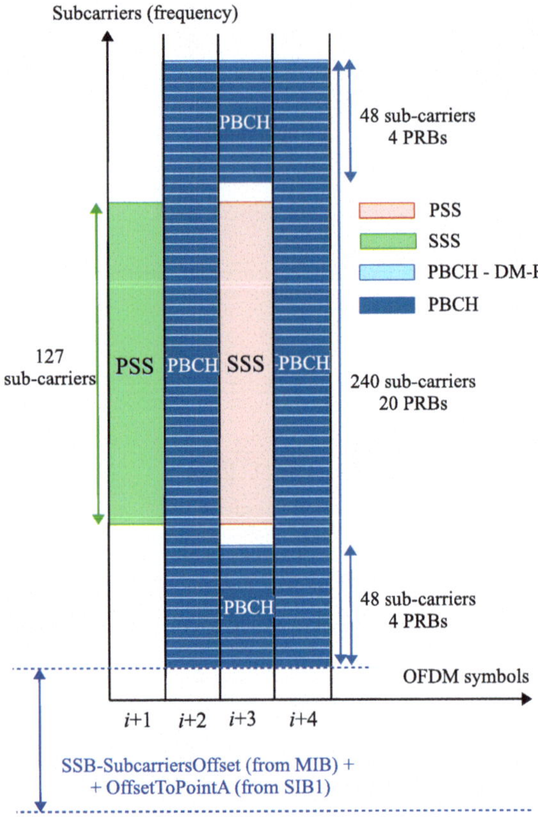

Fig. 1.9 Structure of the Synchronization Signal Block (SSB): 4 consecutive symbols in the time domain and 240 sub-carriers in the frequency domain. The actual duration and bandwidth of the SSB depend on the numerology used

3. The BPCH broadcast signal, delivering some necessary 5G NR system information to a user equipment; in every 4th sub-carrier of the PBCH a Demodulation Reference Signal (DM-RS) signal is sent.

In 5G, PBCH makes use of 20 central Physical Resource Blocks (PRBs) of an OFDM symbol and $20 \cdot 12 = 240$ sub-carriers. The sequence in the SSB block is as follows: PSS (127 sub-carriers), then one complete PBCH symbol, then SSS in the middle, and PBCH around it (4 RBs on both sides), then again one complete PBCH symbol. Together, PSS and SSS define $1008 = 3 \cdot 336$ different physical radio cell identities. After detection and decoding of PSS and SSS using a cross-correlation function, the user equipment, a phone, calculates the physical cell number ($N_{ID}^{cell} = 3N_{ID}^{(1)} + N_{ID}^{(2)}$) and uses it to generate specific pilots exploited for PBCH synchronization. In 5G, all pilot signals are cell- or user-specific. Thanks to them, a phone can achieve downlink time and frequency synchronization with OFDM symbols and acquire time instants of the PBCH channel.

The PBCH carries the basic 5G NR system information necessary to connect to the base station. PBCH has its own DM-RS pilots. It is sent in only one antenna

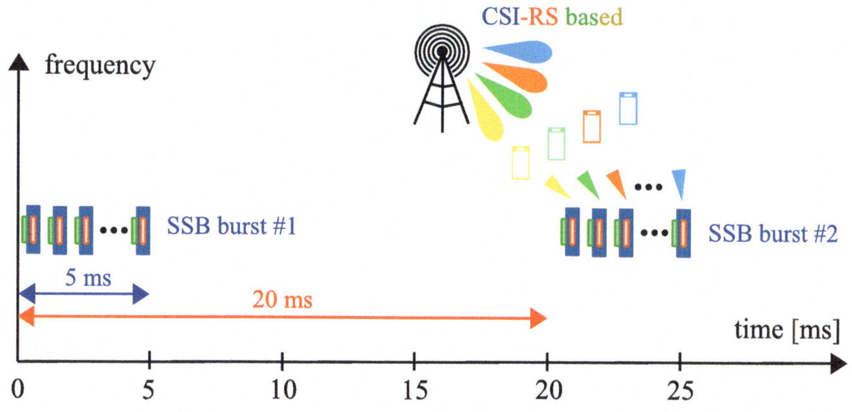

Fig. 1.10 SSB burst set. Each SSB in a burst set is transmitted on the same carrier frequency but with the use of different beams and in different OFDM symbols. Therefore, after detecting the SSB, the UE can also detect the best beam. SSB burst set is repeated every 20 ms

Fig. 1.11 An example of an SSB transmission in a single beam with a 20 ms periodicity. The figure shows a measured signal spectrogram obtained when no downlink data was transmitted. In this case, the only observed transmission consists of SSB bursts periodically sent by the gNB

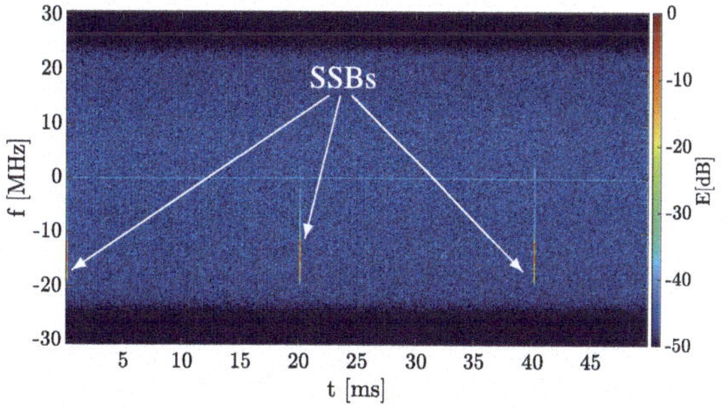

port (no TX diversity is applied, the signal is linearly precoded using beamforming) and coded by polar codes. It is transmitted in bursts several (4, 8, 16) times during a radio frame using different beams—as explained in Fig. 1.10. As a result, a user phone recognizes the best beam for it (i.e., with the highest power) during access to the network and reports its number to the gNB. Then, the gNB uses the selected beam for communication with the device, and the UE uses the antenna with the strongest signal.

SSBs are periodically broadcast in bursts across all beams at intervals $T_{\text{dist}}^{\text{SSB}} \in \{5, 10, 20, 40, 80, 160\}$ ms, depending on the configuration. The default periodicity is 20 ms, but it can be reduced to 10 ms to improve velocity unambiguity in SSB-based sensing. A real-world recording of SSBs transmitted within a single beam is presented in Fig. 1.11.

In separate OFDM time–frequency atoms of any SSB in a burst, signals are sent by only one antenna (port).

SSB Decoding and Reconstruction Process

The signal must be decoded and reconstructed to use the SSB in passive sensing. The first step in the SSB decoding process is the estimation of:

1. $\hat{g}$—the correct Global Synchronization Channel Number (GSCN) position;
2. $\hat{u}$—the PSS identifier $N_{\mathrm{ID}}^{(2)}$;
3. $n_{\mathrm{start}} = \hat{n} - N_{\mathrm{CP}} - N_{\mathrm{FFT}}$—the SSB first sample index, where N_{CP} is the cyclic prefix length in samples and N_{FFT} is the FFT size. The estimation is carried out according to the following expression:

$$\left(\hat{g}, \hat{u}, \hat{n}\right) = \underset{g,u,n}{\operatorname{argmax}} \sum_{m=0}^{M-1} \left| r_u^{\mathrm{PSS}}[m] \cdot x_r^{g*}[m+n] \right|^2, \tag{1.1}$$

where $x_r^g[n]$ is the received signal, shifted to the GSCN raster position $(3000 + g \cdot 1.44)$ MHz, $g \in \{0, 1, ..., 14756\}$—see Sect. 7.4.2.2 in [20], and $(\cdot)^*$ is the complex conjugate, $r_u^{\mathrm{PSS}}[n]$ is the IFFT of $d_u^{\mathrm{PSS}}[n]$, $u \in \{0, 1, 2\}$—the PSS PRBS with $N_{\mathrm{ID}}^{(2)} = u$, see Sect. 7.4.2.2.1 in [15].

By knowing $\hat{g}$, $\hat{u}$, $\hat{n}$, one can obtain the exact time and frequency position of the SSB and the corresponding $N_{\mathrm{ID}}^{(2)}$.

The SSS signal beginning is shifted by two OFDM symbols in the time domain compared with the PSS (see Fig. 1.9) and is generated with the use of Gold sequences as described in Sect. 7.4.2.3.1 in [15]. Its position is already known after the PSS time and frequency synchronization. Therefore, its decoding process is less computationally complex, allowing for the correlation of the FFT weights of the received signal containing SSS ($X^{\mathrm{SSS}}[k]$) with up to 336 different reference sequences denoted as $N_{\mathrm{ID}}^{(1)} \in \{0, 1, ..., 335\}$. The correlation result allows for estimation $\hat{v}$—the SSS identifier $N_{\mathrm{ID}}^{(1)}$ according to

$$N_{\mathrm{ID}}^{(1)} = \hat{v} = \underset{v}{\operatorname{argmax}} \sqrt{\sum_{k=0}^{126} \left| \left(X^{\mathrm{SSS}}[k] d_v^{\mathrm{SSS}*}[k] \right)^2 \right|}, \tag{1.2}$$

where

- $X^{\mathrm{SSS}}[k]$—the SSS FFT weights,
- $d_v^{\mathrm{SSS}}[k]$, $v \in \{0, 1, ..., 355\}$—the SSS PRBS with $N_{\mathrm{ID}}^{(1)} = v$, see Sect. 7.4.2.3.1 in [15].

Estimated IDs of transmitted PSS and SSS sequences are used for the Physical Cell Identifier $N_{\mathrm{ID}}^{\mathrm{cell}}$ calculation as follows:

$$N_{\mathrm{ID}}^{\mathrm{cell}} = 3N_{\mathrm{ID}}^{(1)} + N_{\mathrm{ID}}^{(2)}. \tag{1.3}$$

The PBCH takes two OFDM symbols and one symbol together with the SSS (Fig. 1.9). The resources of all three PBCH OFDM symbols are also mixed with a demodulation reference signal (DM-RS). Since DM-RS is used in different 5G channels in this chapter, the DM-RS of PBCH is denoted as PBCH DM-RS. The DM-RS mapping morphology is cell- and beam-specific and, according to Table 7.4.3.1-1 of [15], it depends on the physical cell identifier $N_{\mathrm{ID}}^{\mathrm{cell}}$, calculated using (1.3), and the SSB/beam identifier i_{SSB}. The value of i_{SSB} is found by calculation of a cross-correlation coefficient between a cut-out PBCH DM-RS $X^{\mathrm{DM\text{-}RS}}[k]$ and the reference PBCH DM-RS PRBS $d_i^{\mathrm{DM\text{-}RS}}$, generated as described in Sect. 7.4.1.4.1 of [15]:

$$i_{\mathrm{SSB}} = \operatorname*{argmax}_i \sqrt{\sum_{k=0}^{144} \left| \left(X^{\mathrm{DM\text{-}RS}}[k] d_i^{\mathrm{DM\text{-}RS}*}[k] \right)^2 \right|}. \tag{1.4}$$

Ultimately, when $N_{\mathrm{ID}}^{\mathrm{cell}}$ and i_{SSB} are known, one can decode a PBCH payload. The possible symbol location $X^{\mathrm{DM\text{-}RS}}[k]$ is specified in Table 7.4.3.1-1 in [15]. Suppose the aim is to obtain the transport block of the broadcast channel (BCH), after $X^{\mathrm{PBCH}}[k]$ QPSK demodulation and descrambling (Sect. 7.3.3 of [15] for more details). In that case, one can apply a BCH error correction and decoding processing chain as described in Sect. 7.1 of [16]. This step ends the SSB decoding procedure. The decoding algorithm is valid for the SSB that can be detected within a given 5G cell.

After the reception and decoding of the SSB, it is possible to reconstruct it into a pure waveform. Firstly, PSS and SSS synchronization signals are appropriately generated using the identified $N_{\mathrm{ID}}^{(2)}$ and $N_{\mathrm{ID}}^{(1)}$. Next, the cell- and beam-specific PBCH DM-RS bits are generated in a 5G time–frequency OFDM grid based on $N_{\mathrm{ID}}^{\mathrm{cell}}$ and i_{SSB}. The identified i_{SSB} is used for imitating the detected beam. If different beams are present, this value can be changed to reconstruct different beams. Then, using the decoded transport block and BCH processing, the desired code block is obtained. The scrambling process results in the PBCH payload. The BPSK modulation is applied for synchronization, reference signals (PSS, SSS, and DM-RS, respectively), and the QPSK modulation for the PBCH payload. Then, all data are mapped to their dedicated positions in the synthesized SSB ($X^{\mathrm{SSB}}[l, k]$, where l is the OFDM symbol and k is the sub-carrier). The precise time instant of the SSB start is given to the ISAC sensing procedure for time synchronization. Figure 1.12 presents the SSB reconstruction process in a typical two-antenna receiver gathering the reference $x_{\mathrm{ref}}[n]$ and surveillance $x_{\mathrm{sur}}[n]$ signal samples. The first one receives the direct signal from the gNB, while the latter collects the reflections from obstacles in the space. This is one of the most commonly used configurations of radar and ISAC systems. More about such an architecture is in the next sections.

Patterns of 5G Reference Signals

As mentioned, 5G reference signals (i.e., pilots) are designed to help receivers in time–frequency synchronization and identify and correct different transmission obstacles. In many configurations, their values and time–frequency positions are precisely defined in the standard [15]. It is only essential for us to remember that time–frequency patterns of reference signals are numerous and can be very different.

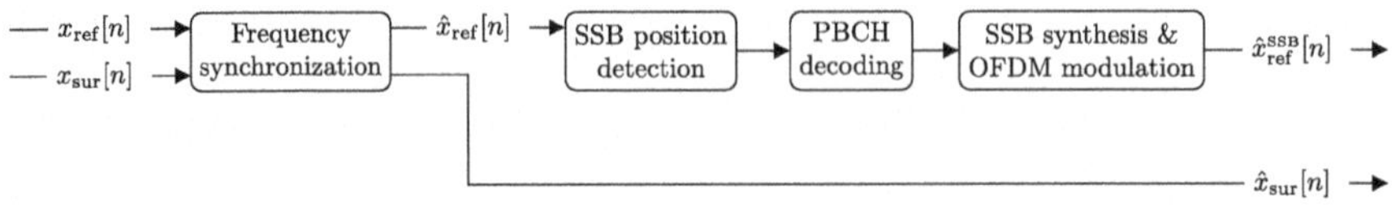

Fig. 1.12 SSB signal reconstruction. The diagram shows the step-by-step processing used to extract a pure SSB reference from a 5G signal. Reproduced from [21] under the CC BY 4.0 license

As an example, in Fig. 1.13, some time–frequency patterns of different 5G reference signals are shown. Generally, the following observations are valid (if the signals are configured/used):

- **DM-RS signals** (used for channel equalization during data demodulation) are dense in frequency and sparse in time.
- **PT-RS signals** (used for phase noise equalization during data demodulation) are dense in time and sparse in frequency.
- **SRS signals** (used for channel sounding in the uplink) are dense in frequency and sparse in time.
- **PRS signals** (used for high-precision object positioning) have a variety of options which are chosen according to requirements.
- **CSI-RS signals** (used for channel observation in the downlink) are sparse in time and frequency; each TX antenna (denoted as A1 and A2 in Fig. 1.13) sends CSI-RS exploiting different sub-carrier (use different *antenna port*) in order to allow the UE measurement of different channel paths; thanks to this efficient pre- and post-coding for spacial multiplexing which can be designed and applied, resulting in better noise robustness of the overall transmission.

1.2.4 Summary

The 5G signal has a strictly defined time–frequency grid structure in the physical layer. Its example is shown in Fig. 1.14. Each cell of this grid represents a DFT coefficient (IQ sub-carrier state). A 5G transmitter sets it into a concrete state, while the receiver tries to recover the sent value. Each cell can generally carry different information: data, control, gNB broadcast, and UE access management. Therefore, each engineer working with the 5G standard should be familiar with the maze/labyrinth of the 5G time–frequency grid settings. In Fig. 1.14, we see: a burst of SSB blocks, a repeating PDCCH control channel, a repeating CSI-RS signal, as well as some extra Radar-on-Demand (RoD) signals, which are not defined in the standard, and can be allocated by the base station for radar purposes as an additional service.

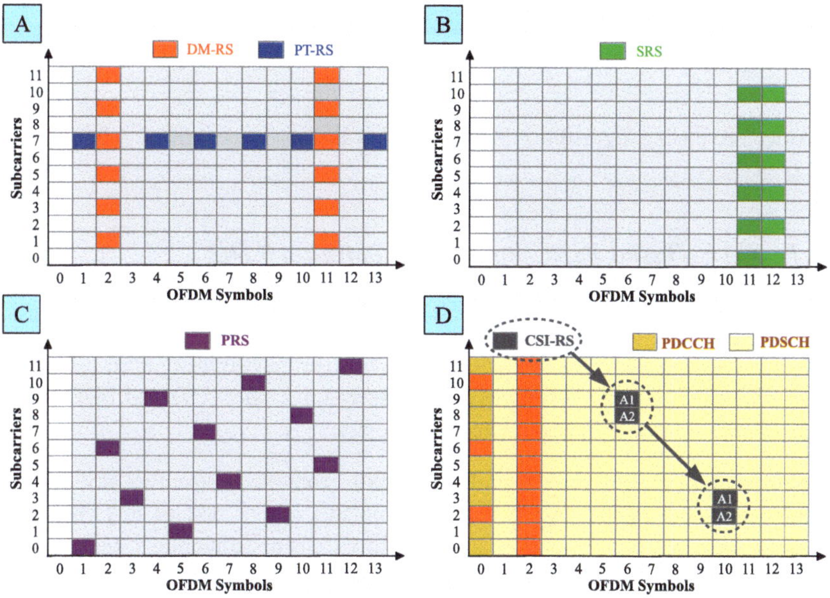

Fig. 1.13 Exemplary time–frequency patterns of different 5G reference signals (RS) used for: demodulation (DM-RS), channel sounding in the uplink (SRS), UE positioning (PRS), and checking downlink channel state information (CSI-RS)

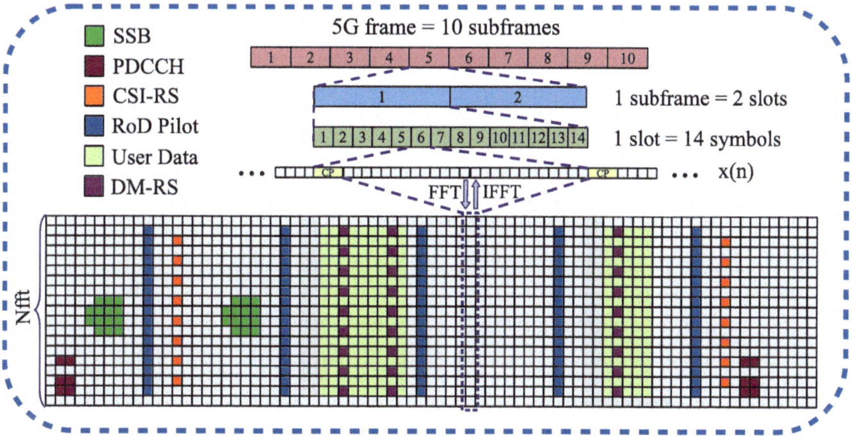

Fig. 1.14 Exemplary 5G time–frequency grid of carriers (vertically) belonging to different OFDM symbols (horizontally) for $\Delta f = 30$ kHz. Reproduced from [22]

1.3 Integrated Sensing and Communication with 5G NR Signal

The selected sensing algorithms dedicated to 5G signals are described in this section.

1.3.1 CAF-Based ISAC for High-Bitstream TDD Transmission

As detailed in this chapter, 5G networks employ two primary full-duplex transmission configurations. From the perspective of radar systems, the TDD and FDD modes exhibit distinctions across several key parameters:

- **Separation of Uplink and Downlink Signals** From the perspective of joint communication and radar sensing (discussed in the further part of this chapter in detail), the separation of uplink and downlink transmission entails the following challenges:
  - **TDD**: Transmission and reception of signals occur at different time instances while utilizing the same frequency band. This necessitates the radar system alternating between transmitting and receiving modes, which can introduce latency and demand precise temporal coordination. Furthermore, sensing equipment must be aware of the prevailing uplink/downlink time allocation ratio.
  - **FDD**: Transmission and reception of signals are carried out concurrently but on distinct frequency bands. This allows for continuous monitoring capabilities and eliminates the requirement for switching between operational modes, which can be advantageous for radar functionalities.
- **Interference and Self-Interference**
  - **TDD**: Given that transmit and receive signals share the same frequency spectrum, there exists a potential for interference between these signals in the radar subsystem if the switching process is not perfectly synchronized. This necessitates the implementation of sophisticated techniques for self-interference cancellation.
  - **FDD**: The inherent frequency separation between transmit and receive signals in FDD mitigates the risk of interference. However, this approach necessitates filtering mechanisms and can potentially increase the cost and complexity of the overall system.
- **Delays and Response Time**
  - **TDD**: The inherent switching between transmission and reception in TDD can introduce temporal delays, which may impact the radar's response time and capability to effectively track rapidly moving targets.

  - **FDD**: FDD enables simultaneous transmission and reception, thereby minimizing delays and potentially enhancing the radar's ability to track dynamic targets with greater accuracy.

- **Spectrum Management**
  - **TDD**: Requires fewer frequency resources as transmission and reception utilize the same frequency, albeit at different points in time.
  - **FDD**: Necessitates the division of the available spectrum into two separate frequency bands, which might be less efficient from a spectrum utilization standpoint but offers improved signal quality by minimizing interference.

- **Applications**
  - **TDD**: Frequently employed in radar systems where stringent real-time monitoring requirements are less critical, such as environmental monitoring systems or meteorological radars.
  - **FDD**: Preferred in applications that demand rapid response times and high tracking precision, such as military radar systems or aviation surveillance radars.

The TDD mode is the focus of further investigation. In essence, TDD facilitates both uplink and downlink data transmission on an identical carrier frequency by allocating distinct time intervals for each direction. The temporal proportion dedicated to uplink and downlink within a frame can vary depending on the specific requirements of a given base station, and the complete distribution of all possible ratios is illustrated in Fig. 1.15.

Observing the grid, it becomes evident that a greater allocation of symbols to downlink transmission is advantageous for sensing performance. It is important to note the high degree of flexibility inherent in 5G networks, exemplified by formats such as 16–18, where up to 13 symbols within a slot can be dynamically configured as either uplink or downlink, contingent upon specific demands. This adaptability suggests the potential for managing and designing the network according to precise operational needs. However, leveraging the 5G signal for radar applications under TDD necessitates careful consideration of its time-multiplexed nature.

Conversely, in the case of FDD, all symbols within a slot allocated to a downlink carrier are exclusively used for downlink transmissions. Similarly, all symbols within an uplink carrier's slot are dedicated solely to uplink transmissions. The sensing aspect is considerably simplified in FDD as it does not require differentiation between signal directions over time. Nevertheless, dividing the available bandwidth into two separate segments results in a range resolution twice as coarse as TDD for the same total bandwidth. This is why the TDD mode has better performance in terms of radar operation, but switching between downlink and uplink segments is more challenging.

Signal processing methods for 5G-based sensing part in ISAC systems The various subsections detailed in the subsequent parts of this chapter can be considered

Format	Symbol number in a slot													
	0	1	2	3	4	5	6	7	8	9	10	11	12	13
0	D	D	D	D	D	D	D	D	D	D	D	D	D	D
1	U	U	U	U	U	U	U	U	U	U	U	U	U	U
2	F	F	F	F	F	F	F	F	F	F	F	F	F	F
3	D	D	D	D	D	D	D	D	D	D	D	D	D	F
4	D	D	D	D	D	D	D	D	D	D	D	D	F	F
5	D	D	D	D	D	D	D	D	D	D	D	F	F	F
6	D	D	D	D	D	D	D	D	D	D	F	F	F	F
7	D	D	D	D	D	D	D	D	D	F	F	F	F	F
8	F	F	F	F	F	F	F	F	F	F	F	F	F	U
9	F	F	F	F	F	F	F	F	F	F	F	F	U	U
10	F	U	U	U	U	U	U	U	U	U	U	U	U	U
11	F	F	U	U	U	U	U	U	U	U	U	U	U	U
12	F	F	F	U	U	U	U	U	U	U	U	U	U	U
13	F	F	F	F	U	U	U	U	U	U	U	U	U	U
14	F	F	F	F	F	U	U	U	U	U	U	U	U	U
15	F	F	F	F	F	F	U	U	U	U	U	U	U	U
16	D	F	F	F	F	F	F	F	F	F	F	F	F	F
17	D	D	F	F	F	F	F	F	F	F	F	F	F	F
18	D	D	D	F	F	F	F	F	F	F	F	F	F	F
19	D	F	F	F	F	F	F	F	F	F	F	F	F	U
20	D	D	F	F	F	F	F	F	F	F	F	F	F	U
21	D	D	D	F	F	F	F	F	F	F	F	F	F	U
22	D	F	F	F	F	F	F	F	F	F	F	F	U	U
23	D	D	F	F	F	F	F	F	F	F	F	F	U	U
24	D	D	D	F	F	F	F	F	F	F	F	F	U	U
25	D	F	F	F	F	F	F	F	F	F	F	U	U	U
26	D	D	F	F	F	F	F	F	F	F	F	U	U	U
27	D	D	D	F	F	F	F	F	F	F	F	U	U	U
28	D	D	D	D	D	D	D	D	D	D	D	D	F	U
29	D	D	D	D	D	D	D	D	D	D	D	F	F	U
30	D	D	D	D	D	D	D	D	D	D	F	F	F	U
31	D	D	D	D	D	D	D	D	D	D	D	F	U	U
32	D	D	D	D	D	D	D	D	D	D	F	F	U	U
33	D	D	D	D	D	D	D	D	D	F	F	F	U	U
34	D	F	U	U	U	U	U	U	U	U	U	U	U	U
35	D	D	F	U	U	U	U	U	U	U	U	U	U	U
36	D	D	D	F	U	U	U	U	U	U	U	U	U	U
37	D	F	F	U	U	U	U	U	U	U	U	U	U	U
38	D	D	F	F	U	U	U	U	U	U	U	U	U	U
39	D	D	D	F	F	U	U	U	U	U	U	U	U	U
40	D	F	F	F	U	U	U	U	U	U	U	U	U	U
41	D	D	F	F	F	U	U	U	U	U	U	U	U	U
42	D	D	D	F	F	F	U	U	U	U	U	U	U	U
43	D	D	D	D	D	D	D	D	D	F	F	F	F	U
44	D	D	D	D	D	D	F	F	F	F	F	F	U	U
45	D	D	D	D	D	D	F	F	U	U	U	U	U	U
46	D	D	D	D	D	F	U	D	D	D	D	D	F	U
47	D	D	F	U	U	U	U	D	D	F	U	U	U	U
48	D	F	U	U	U	U	U	D	F	U	U	U	U	U
49	D	D	D	D	F	F	U	D	D	D	D	F	F	U
50	D	D	F	F	U	U	U	D	D	F	F	U	U	U
51	D	F	F	U	U	U	U	D	F	F	U	U	U	U
52	D	F	F	F	F	F	U	D	F	F	F	F	F	U
53	D	D	F	F	F	F	U	D	D	F	F	F	F	U
54	F	F	F	F	F	F	F	D	D	D	D	D	D	D
55	D	D	F	F	F	U	U	U	D	D	D	D	D	D
56 – 254	Reserved													
255	UE determines the slot format for the slot based on *tdd-UL-DL-ConfigurationCommon*, or *tdd-UL-DL-ConfigurationDedicated* and, if any, on detected DCI formats													

Fig. 1.15 All possible uplink/downlink time resource divisions in 5G networks. The table illustrates how the 5G standard is flexible allowing multiple configurations for various applications. Downlink is represented by the symbol D, uplink by U, and flexible allocation by F. *Source* [17], Table 11.1.1-1. Reproduced from [23]

(a) Active 5G-based sensing scenario (b) Passive 5G-based sensing scenario

Fig. 1.16 Two scenarios of 5G-based sensing. In scenario **a**, the base station simultaneously transmits the communication signal and receives its echoes. In scenario **b**, the communication signal is exploited by an external ISAC system for target detection. Reproduced from [2]

within the context of two distinct sensing scenarios. The primary scenario involves the sensing functionality integrated directly into the 5G base station. In this configuration, the base station transmits and receives signals for the purpose of communication and, concurrently, estimates the position and velocity of targets within its vicinity based on the received signal. This approach presents a relatively less complex sensing perspective, as the signal is *transparent* to the base station. The station can fully synchronize, demodulate, and decode the signal. This transparency results in a higher signal-to-noise ratio (SNR) following sensing processing and streamlines the overall processing pipeline due to the base station's complete control over the radio channel. The secondary scenario pertains to a non-cooperative and bistatic configuration, wherein the receiver is spatially separated from the base station and lacks comprehensive control over the transmitted signal.

These two distinct scenarios are visually represented in Fig. 1.16 [2]. These configurations are critical in several aspects of ISAC signal processing.

The fundamental concept of radar signal processing in 5G-based ISAC for the monostatic and cooperative case (Option 1) and the bistatic&passive radar cases (Options 2, 3, 4) is depicted in Fig. 1.17 [2]. In Option 1, the base station operates cooperatively with a radar receiver, where its functionality can be integrated within the base station. Conversely, Options 2 and 4 utilize non-cooperative illumination sources. Specifically, in Option 2, which can be characterized as a bistatic and non-cooperative operation, the entire full-content signal acquired from a reference antenna is employed as a template. Subsequently, in TDD mode, the uplink portion is additionally removed from this template. In Options 3 and 4, the reconstruction of the 5G synchronization block and control signals is leveraged for matched filtering. These components are extracted from the reference antenna (Option 3) or the surveillance antenna (Option 4). Option 4 enables the realization of a single-antenna bistatic&passive radar system. However, operation in Option 3 necessitates the availability of at least two receiving antennas.

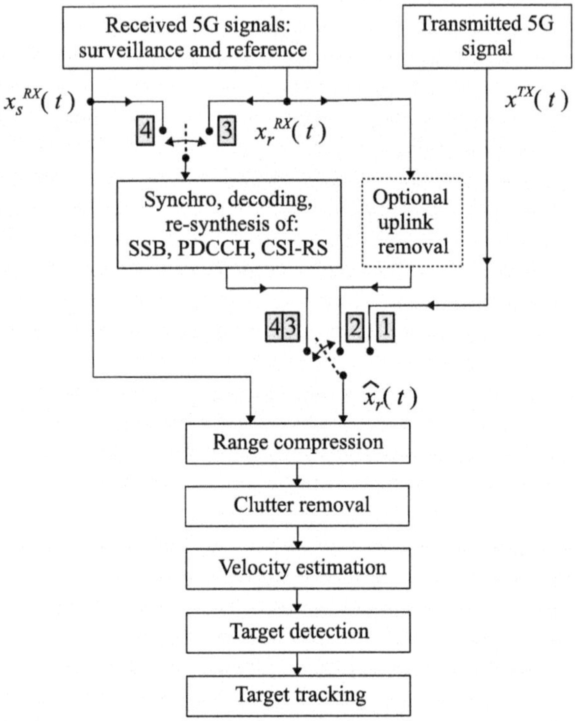

Fig. 1.17 A general signal processing diagram for radar using a 5G network as a source of illumination. Depending on the configuration, the 5G signal can be exploited for active and passive sensing with communication. Reproduced from [2]

As previously mentioned, the signal processing methodologies for both types of radar configurations, namely, monostatic&cooperative and bistatic&passive, exhibit significant similarities. In the latter (bistatic&passive) case, the received signal can be additionally synchronized, which leads to an enhancement of the SNR at the output of the matched filter. It is important to note that this synchronization operation is not always strictly obligatory. The process of range compression is performed using either matched filtering (for monostatic&cooperative radar) or correlation techniques (for bistatic&passive radar). The subsequent processing stages involve clutter removal, velocity estimation of targets, target detection, and the tracking of detected targets. The underlying principles governing the processing of each of these options are elaborated upon in the following subsections of this chapter.

1. **FDD**
   FDD is characterized by its continuous operation and separation of uplink and downlink transmissions through distinct carrier frequencies. Consequently, both data transmission and sensing functionalities can be performed concurrently within their respective frequency bands. Data demodulation and decoding are executed in accordance with the standard and the physical layer specifications outlined in this chapter. From a sensing perspective, the 5G signal exhibits noise-like characteristics due to its utilization of OFDM for data transmission. In radar

applications, the OFDM signal may or may not be reconstructed, depending on the cooperative nature of the transmitting gNB. However, irrespective of this reconstruction step, signal processing for target detection (in particular, target range and velocity estimation) can be based on the cross-ambiguity function (CAF) [24, 25]:

$$\Psi(R, V) = \int_0^{T_i} x_s(t)\, x_r^*\left(t - \frac{R}{c}\right) \exp\left(-j\frac{2\pi}{\lambda} V t\right) \mathrm{d}t, \tag{1.5}$$

where T_i represents the integration time, R denotes the target range, c is the speed of light, V signifies the target velocity, and λ is the wavelength of the signal. The parameters R and V can correspond to a monostatic configuration if the 5G gNB or user equipment operates as a cooperative active radar system and to a bistatic configuration when the receiver is non-cooperative and/or spatially separated from the transmitter (as depicted in Fig. 1.16). The signal processing pathway for the FDD mode can be identified as Option 1 or 2 in Fig. 1.17. Notably, for Option 2 in the FDD context, the block labeled "*Optional uplink removal*" is not applicable, as the uplink signal does not interfere with the downlink signal due to its inherent frequency separation. Option 1 represents a classical active sensing scenario where two signals, a reference $x_r(t)$ and a surveillance signal $x_s(t)$, are correlated according to (1.5). The passive sensing system (Option 2) operates analogously, with the key difference being that the reference signal is not known *a priori*. This constitutes the most direct approach to radar signal processing within 5G-based ISAC systems employing FDD.

2. **TDD**

   5G-based ISAC in TDD mode presents greater challenges than FDD due to the time-multiplexed nature of uplink and downlink transmissions on the same carrier frequency, making precise time synchronization critical for radar performance. Key difficulties from a radar perspective include [1, 2]:

   - A radar receiver, especially a non-cooperative one, needs network numerology details (TDD pattern, intervals, frequencies, etc.) for synchronization, which is often unavailable, hindering signal processing.
   - Identifying and handling uplink signals from UEs is necessary. These low-power signals can cause false or ambiguous target detections if not effectively removed (e.g., by zeroing or filling uplink slots with noise).
   - Removing uplink transmissions reduces the effective integration time, degrading bistatic velocity resolution [25] and decreasing integration gain. Proper T_i selection is also limited by range and velocity cell migration, constraining coherent processing time [25]. Despite reduced effective T_i, the total number of samples remains the same, a factor in real-time radar design for sufficient bistatic velocity resolution.
   - While short 5G NR integration times allow frequent measurement updates (e.g., 20 Hz for $T_i = 50$ ms), real-time processing of the large data volume at

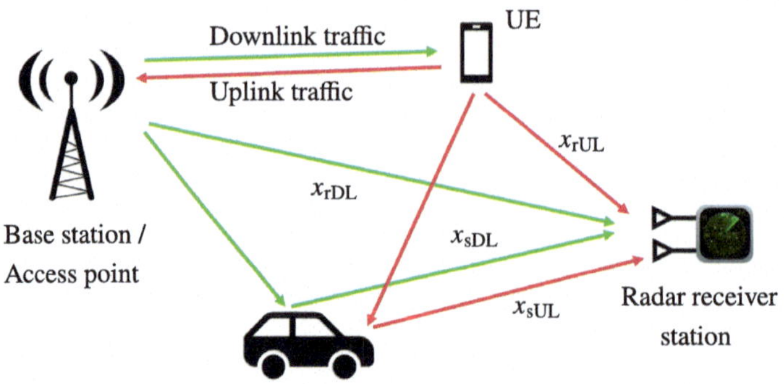

Fig. 1.18 Geometry of the sensing component in an ISAC system based on TDD transmission. Green arrows indicate signals transmitted by the base station/access point (downlink), while red arrows indicate signals transmitted by the user equipment (uplink). This geometry may introduce additional echoes in radar analysis, as the UE acts as an extra illumination source for the target. Reproduced from [23]

Fig. 1.19 Visualization of the additional illumination effect. A false target appears when a moving object is illuminated by the UE

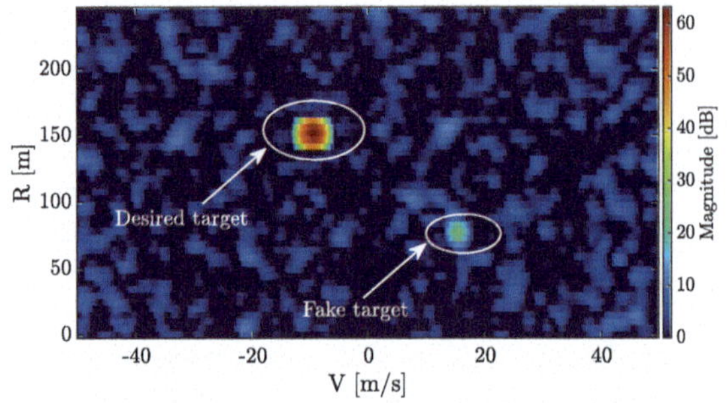

high bandwidths can exceed this time. Skipping data to achieve lower refresh rates (e.g., a few Hz) is suggested for applications like car or UAV monitoring to facilitate real-time operation.

In TDD mode, transmission on a single frequency band is bidirectional and time-switched based on a network-defined schedule [23], altering the passive radar geometry (Fig. 1.18). Green arrows indicate downlink (DL) signals from the gNB, used by UEs for communication and received directly by the radar (x_{rDL}). The reflected downlink signal (x_{sDL}) arrives with delay and Doppler, enabling target detection via CAF. Conversely, orange arrows represent uplink signals from UEs. The reflected uplink signal also reaches the radar with its delay and Doppler, causing interfering detections in the CAF when correlated with the direct downlink signal (see Fig. 1.18).

The simulated effect of two illumination sources (when the first—base station—is desired and the second one—user terminal/UE—is unwanted) on the target

sensing capability can be seen in Fig. 1.19. The first, desired target, is apparent at 150 m and has a velocity of -10 m/s. The echo originates from the reflection of the 5G signal from a moving target. Additionally, the UE illuminated the same space, creating a second geometry according to Fig. 1.18. The fake echo ($\times 100$ weaker) can be seen at 75 m with a velocity of 15 m/s.
To avoid uplink–downlink interference in the sensing part when the TDD mode is used, one can remove the signal from the UE (uplink) [1]. This step is computationally complex but can be required for a strong uplink signal, e.g., when the sensing part is located close to the UE. Blocks of the processing pipeline affected by that work were marked red in Fig. 1.20.
Unlike traditional sensing without reference signal reconstruction, uplink removal in 5G TDD ISAC necessitates time and frequency synchronization and 5G network configuration knowledge. The general uplink removal process [1] involves

(a) Initialization: Configuring 5G network parameters (sampling rate, OFDM symbol/CP lengths, sub-carriers, spacing).
(b) Signal resampling: Matching transmitter and receiver sampling rates.
(c) Self-correlation: Locating OFDM symbol starts using cyclic prefix correlation.
(d) Fractional carrier frequency offset estimation: Fine-tuning frequency synchronization.
(e) Synchronization signal generation: Creating three known synchronization sequences.
(f) Synchronization sequence identification: Finding the best correlation with the received signal.
(g) Frequency offset compensation: Modulating received signals ($x_r(t)$ and $x_s(t)$) based on the identified frequency offset.
(h) Uplink removal: Replacing samples corresponding to uplink transmissions (based on Fig. 1.15 and network configuration) with zeros.

The result of the uplink signal cancellation can be seen in Fig. 1.21. This example shows the real-life signal recorded from the cooperative 5G network. The uplink transmission (although significantly weaker than the downlink signal) is apparent as bright sequences on the spectrogram from Fig. 1.21a. After uplink cancellation, the slots containing this signal from the user equipment were replaced by zeros. The effect is apparent in Fig. 1.21b.

3. **Content dependency**: Regardless of the transmission mode (TDD/FDD), 5G networks transmit signals on demand, emitting downlink signals only when users request content. This approach aligns with electromagnetic smog reduction and energy efficiency. However, the variability in downlink signal duration and bandwidth leads to inconsistent target sensing performance for passive receivers. Specifically, utilizing only a portion of the available time and frequency resources degrades range and velocity resolution and the target SNR (1.5). The impact of this content-dependent transmission on sensing is illustrated in Fig. 1.22. The results clearly show that varying content can significantly affect sensing capabilities. When no user downloads data, the base station periodically transmits

synchronization blocks (in this instance, every 20 ms, visible as a short pulse between −20 and −10 MHz of the normalized bandwidth at the beginning of the sequence).

From the perspective of passive utilization of the 5G signal for sensing, time slots when the base station does not transmit any signal (or a very small signal), there is no reason to perform processing (1.5). Thus, in [23], the adaptive sensing method was proposed. The processing is based on the spectrogram computed using the short-time Fourier transform (STFT), which for the complex and continuous signal $x(t)$ and the analyzing window $h(t)$ is defined as [26]

$$F_x^h(t,\omega) = \int_{\mathbb{R}} x(\tau)h(\tau - t)\mathrm{e}^{-j\omega(\tau-t)}\mathrm{d}\tau. \tag{1.6}$$

The energy distribution (spectrogram) is given as a squared modulus of (1.6) and defined as

$$S_x^h(t,\omega) = |F_x^h(t,\omega)|^2. \tag{1.7}$$

The algorithm takes the reference signal $x_r(t)$ as the input for the STFT computation.

The mathematical representation of the OFDM signal is formulated as follows [27]:

$$x(t) = \frac{1}{\sqrt{N}} \sum_{k=0}^{N-1} d_k e^{j2\pi f_k t}, \quad 0 < t < T, \tag{1.8}$$

where f_k denotes the frequency of the kth sub-carrier selected from a group of N orthogonal sub-carriers, d_k is the complex-valued data symbol mapped to the kth sub-carrier, and T corresponds to the duration of a single OFDM symbol. The expression in (1.8) applies to 5G communication systems and effectively represents the inverse discrete Fourier transform applied to the modulated sub-carriers. Consequently, the STFT defined in Eq. (1.6) can be interpreted as performing the reverse operation of OFDM. In this context, the spectrogram illustrates the distribution of sub-carriers in both time and frequency domains simultaneously. In 5G networks, the allocation and duration of sub-carriers dynamically vary in response to instantaneous data demands. The resulting time–frequency representation provides insight into how data resources are utilized during transmission. For the general OFDM formulation (1.8) and the corresponding spectrogram (1.7), information-theoretic tools can be employed to evaluate resource utilization in the time–frequency domain. One such measure is the Rényi entropy, which quantifies the information content of the spectrogram (1.7) in bits. It is defined as follows [28]:

$$H_R^{\gamma}(\mathrm{TF}_x) = \frac{1}{1-\gamma} \log_2 \left(\frac{\iint_{\mathbb{R}^2} (\mathrm{TF}_x(t,\omega))^{\gamma}\, \mathrm{d}t\mathrm{d}\omega}{\iint_{\mathbb{R}^2} (\mathrm{TF}_x(t,\omega))\, \mathrm{d}t\mathrm{d}\omega} \right), \tag{1.9}$$

where $\mathrm{TF}_x(t, \omega)$ denotes a time–frequency representation of the signal (in this study, the spectrogram given in (1.7)), and γ is the order of the entropy. This work adopts $\gamma = 3$, as it yields stable outcomes and is recommended for effective information content analysis [28].

High variability characterizes 5G downlink content. Therefore, Rényi entropy should be analyzed over short time slots (e.g., 20 ms) using (1.9) to capture these changes. Passive sensing subsystems might need calibration via intentional maximum time and bandwidth allocation (e.g., by downloading large files) to determine the maximum entropy value, representing full resource utilization. This allows for adaptive integration processing, where only signal segments with entropy exceeding a defined threshold are used for subsequent processing. The signal processing flow for this adaptive, content-dependent transmission approach is summarized in Fig. 1.23 [23, 29].

The diagram illustrates that the 5G-based sensing framework presented in Fig. 1.20 is enhanced by incorporating additional stages: spectrogram generation, Rényi entropy computation, and a decision process for integration. Processing is initiated only if the computed entropy surpasses a predefined threshold. Otherwise, the signal content is considered insufficient for reliable target detection. An example of a real-life 5G signal spanning 1 s, with entropy evaluated over consecutive 20 ms segments, is depicted in Fig. 1.24. The results indicate that the entropy value decreases in the absence of downlink transmissions. In contrast, during periods of intense data transmission, the entropy rises, enabling adaptive signal integration strategies [23, 29].

Based on this observation, one might conclude that sensing is unfeasible when the 5G ISAC signal lacks user data content. However, this is not entirely accurate. Even without its own link payload, the gNB periodically transmits synchronization signals to maintain time and frequency alignment within the network. The use of these signals for sensing applications is discussed in the following subsection.

1.3.2 CAF-Based ISAC for Low-Bitstream TDD&FDD Transmission Using SSB

In 5G networks, content-dependent transmission can limit the feasibility of sensing, especially in rural or border areas with low user density. These regions are ideal for short-range passive sensing and ISAC applications such as detecting vehicles, individuals, or smuggler drones. Base stations in such areas are typically sparsely deployed, often operating at lower carrier frequencies to improve coverage through forests and foliage. They also tend to transmit at higher power but with lower traffic than urban networks. Another relevant scenario involves drone detection near small or aeroclub-operated airports.

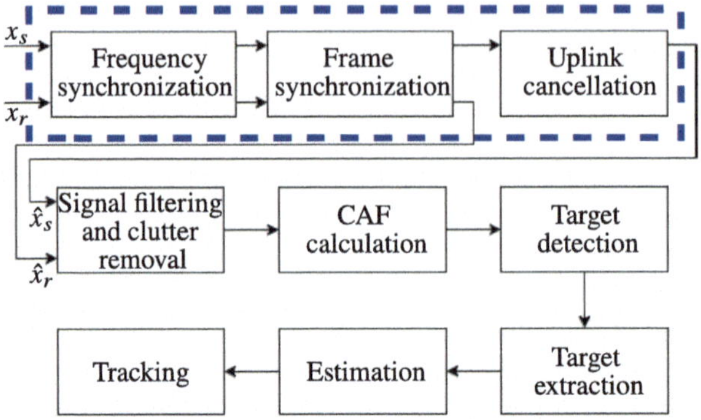

Fig. 1.20 An example of a signal processing concept for an ISAC system based on a 5G waveform. The processing pipeline differs from classical approaches by introducing additional steps, including uplink signal cancellation for false echo removal. The blue box indicates the additional processing stage. Reproduced from [1]

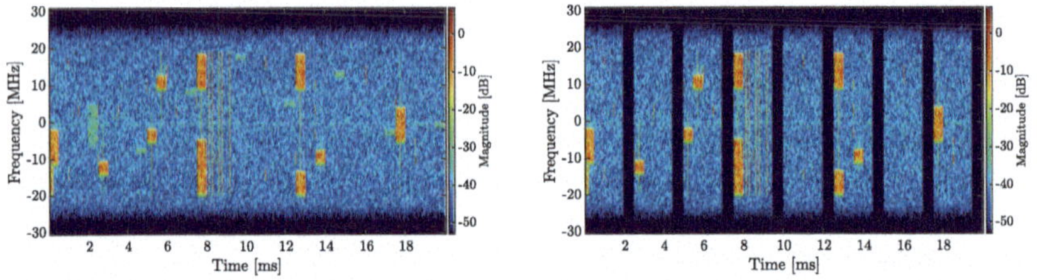

(a) Spectrogram of the real-life 5G signal before uplink signal removal

(b) Spectrogram of the real-life 5G signal after uplink signal removal

Fig. 1.21 Visualization of uplink signal removal in a 5G-based sensing system. Since uplink and downlink transmissions share the same carrier frequency, the uplink component must be removed to prevent interference and false targets. Reproduced from [1]

Importantly, 5G base stations transmit the SSB regardless of user activity. The SSB is the only always-on signal in a 5G system and includes a broadcast channel that must be detectable within the entire coverage area. When beamforming is used, multiple beam-specific SSBs are transmitted, forming a Synchronization Signal Burst (SS Burst) [30]. In the absence of user data, the SSB remains a viable source for both active (base station) and passive (receiver) sensing [21].

Sensing is performed in a bistatic, non-cooperative ISAC setup using two antennas. The reference signal $x_{\text{ref}}[n]$ is captured from the gNB direction, while the surveillance signal $x_{\text{sur}}[n]$ monitors the target area. Both signals are first resampled and synchronized. SSB reconstruction is then performed only on the reference channel. The SSB is detected using (1.1), followed by PBCH decoding. The time-domain SSB is reconstructed as

$$x^{\text{SSB}}[n] = [\text{CP}(\text{IFFT}(X^{\text{SSB}}[0, :])), ..., \text{CP}(\text{IFFT}(X^{\text{SSB}}[3, :]))], \tag{1.10}$$

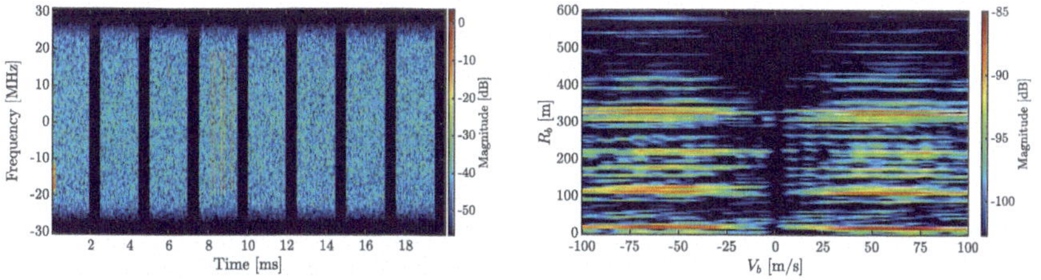

(a) Spectrogram of the 5G signal without transmitted content (b) Range-velocity map for the signal without transmitted content

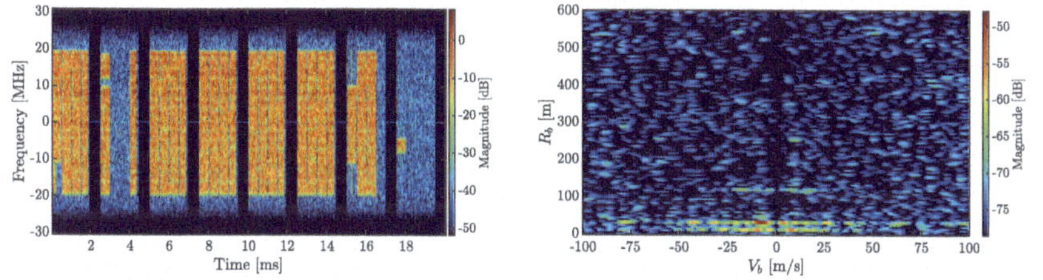

(c) Spectrogram of the 5G signal with a lot of transmitted content (d) Range-velocity map for the 5G signal with a lot of transmitted content

Fig. 1.22 Visualization of the sensing performance depending on the content transmitted by the 5G gNB. Target detection is possible only in the presence of an illuminating signal; when the gNB does not transmit, the passive radar remains blind. Reproduced from [1]

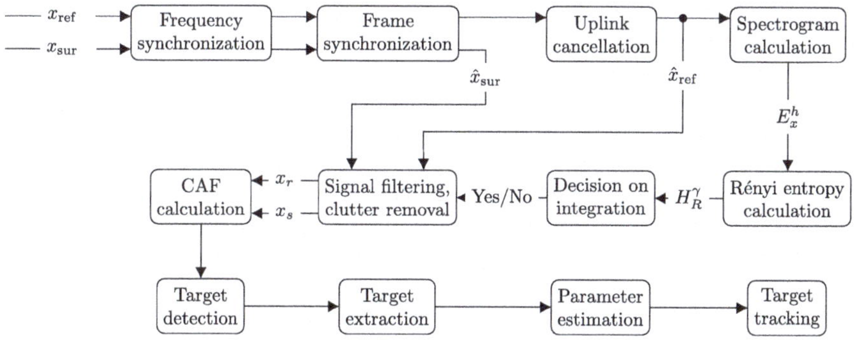

Fig. 1.23 5G-based passive sensing scheme for adaptive, content-dependent downlink signal integration. In this configuration, the algorithm decides on signal integration based on the downlink content and the estimated entropy. Reproduced from [29] under the CC BY 4.0 license

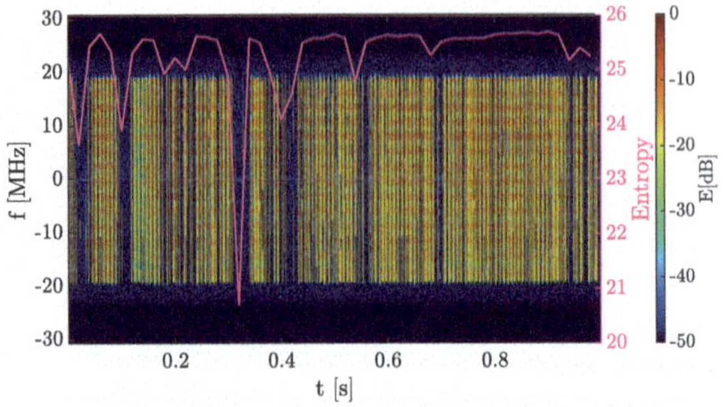

Fig. 1.24 Spectrogram of a real 5G signal with Rényi entropy computed over 20-s intervals. The estimated Rényi entropy drops sharply whenever the gNB ceases downlink transmission. Reproduced from [29] under the CC BY 4.0 license

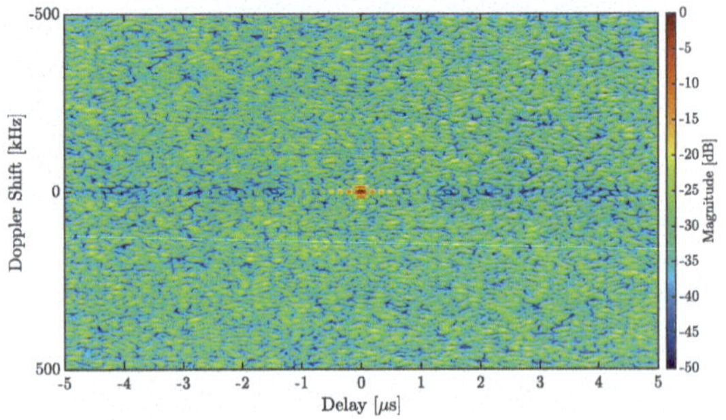

Fig. 1.25 Ambiguity function of the 5G SSB waveform. The signal exhibits favorable correlation characteristics, with a well-focused main peak and low sidelobes

where CP(·) adds a cyclic prefix, and IFFT(·) denotes the inverse fast Fourier transform.

This process yields a clean SSB reference for further sensing. While a single-antenna setup using x_{ref} can support matched filtering, the dual-channel configuration improves detection. SSB reconstruction is unnecessary in cooperative radar integrated within the gNB, as the waveform is already known. This method applies specifically to non-cooperative passive scenarios.

In 5G SSB-based ISAC, SSBs can support moving target detection when classical processing using downlink signals (1.5) is ineffective. Due to its noise-like nature, the SSB exhibits favorable radar characteristics. Ideally, the ambiguity function should have a sharp main peak at zero delay and Doppler, with minimal sidelobes. As shown in Fig. 1.25, the SSB waveform achieves approximately ~30 dB of dynamic range, comparable to pulse radar signals [21].

For passive sensing, the SSB must be reconstructed. It can be done as described in this chapter in Sect. 1.2.3. A range-time matrix is then created, where the fast time axis corresponds to the range (N samples), and the slow time to successive SSBs spaced by $T_{\text{dist}}^{\text{SSB}}$. The number of SSBs processed is $M = \lfloor T_i / T_{\text{dist}}^{\text{SSB}} \rfloor$, where T_i is the integration time. The range-time map is formed by matched filtering (convolution) of the surveillance signal $\hat{x}_{\text{sur}}[n]$ with the time-reversed, conjugated SSB reference $h[n] = \hat{x}_{\text{ref}}^{\text{SSB}}[-n]^*$:

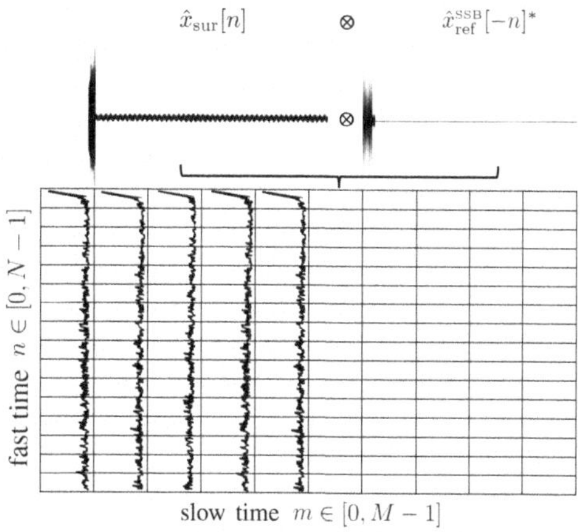

Fig. 1.26 Concept of range-time map formation. The received signal is processed with a matched filter. Similar to classical pulse radar, the interval between two consecutive SSB bursts corresponds to the listening period for target echoes. Reproduced from [21] under the CC BY 4.0 license

$$\mathcal{R}[n, m] = \sum_{i=0}^{N-1} h[i, m]\hat{x}_{\text{sur}}[n - i, m], \tag{1.11}$$

efficiently implemented via FFT:

$$y[n] = \sum_{i=0}^{N-1} h[i]\hat{x}_{\text{sur}}[n - i]. \tag{1.12}$$

Next, phase alignment compensates for random phase shifts between SSBs

$$\mathcal{R}_a[n, m] = \mathcal{R}[n, m] \cdot \exp\left(-j\hat{\phi}[m]\right), \tag{1.13}$$

where $\hat{\phi}[m] = \measuredangle\left(\mathcal{R}[0, m]\right)$. Clutter is removed by applying a high-pass filter along the slow time

$$\mathcal{R}_a^c[n, m] = \sum_{j=0}^{M-1} w[j]\mathcal{R}_a[n, m - j], \tag{1.14}$$

with $w[j]$ denoting the filter coefficients. A Doppler spectrum is then computed for each range cell using the FFT

$$C[n, k] = \mathcal{F}_{m \to k}\{\mathcal{R}_a^c[n, m]\}, \tag{1.15}$$

where k represents Doppler bins (V_b). Target detection and estimation follow before tracking. The full signal processing chain is summarized in Fig. 1.27. SSB reconstruction and phase alignment are unnecessary in active sensing scenarios, as the transmitter and receiver are co-located.

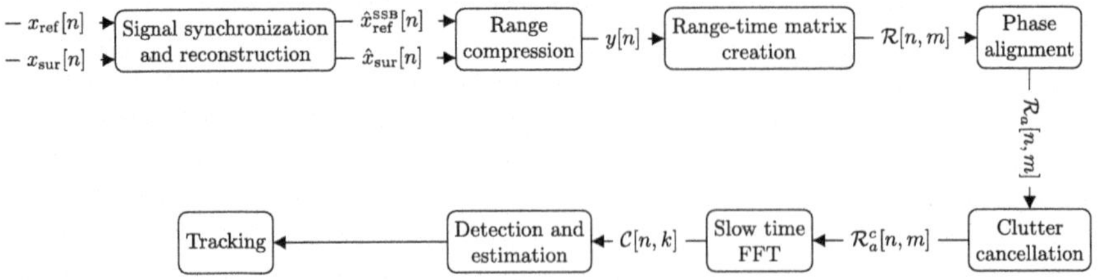

Fig. 1.27 Signal processing pipeline for 5G PCL based on SSB. Apart from SSB synchronization and reconstruction, the processing follows the principles of classical pulse radar. Reproduced from [21] under the CC BY 4.0 license

Table 1.3 Unambiguous velocity (in m/s) for different carrier frequencies f_c and SSB periodicity T^{SSB}_{dist}. The illustrated values represent the limits of maximum target velocities that can be unambiguously estimated in the sensing component of the 5G-based ISAC system

	T^{SSB}_{dist}					
f$_c$ [MHz]	**5**	**10**	**20**	**40**	**80**	**160**
750	±20	±10	±5	±2.50	±1.25	±0.63
3440	±4.36	±2.18	±1.09	±0.54	±0.27	±0.13
5900	±5.17	±2.59	±1.29	±0.64	±0.32	±0.16
25000	±0.60	±0.30	±0.15	±0.08	±0.04	±0.02
60000	±0.25	±0.13	±0.06	±0.03	±0.02	±0.01

The SSBs are transmitted relatively rarely. The Doppler shift for the monostatic geometry for a given target velocity V and wavelength λ is given as

$$f_d = \frac{2V}{\lambda}. \tag{1.16}$$

Thus, the velocity span is limited by T^{SSB}_{dist} so that

$$V \in \left[\frac{-\lambda}{4T^{SSB}_{dist}}, \frac{\lambda}{4T^{SSB}_{dist}}\right). \tag{1.17}$$

This property entails a significant limitation on the unambiguous velocity, which cannot be sufficient for fast and maneuvering target detection. For the monostatic geometry, the unambiguous velocities (in [m/s]) for different carrier frequencies and SSB periodicities are shown in Table 1.3. For the bistatic geometry, the velocities are as large as twice due to the vanishing of 2 in the numerator of (1.17). The values show that, from the perspective of sensing performance, the best option is to operate at the lowest possible carrier frequencies with the highest SSB periodicity. Operating at high frequencies (mmWave) and SSB periodicity extension leads to unambiguity that cannot be accepted in a sensing system.

When the ambiguity is not very high, one can improve sensing performance under some prerequisites. In general, the problem arises when a moving target introduces a frequency shift in the received signal, estimated using a Fourier transform along the slow time (see Figs. 1.27 and 1.26). Therefore, the pulse repetition rate (SSB periodicity) corresponds to the sampling rate in the Doppler frequency domain. According to the Nyquist sampling theorem, the higher the pulse repetition rate, the wider the unambiguous velocity range. As can be deduced, assuming the default value of $T_{\text{dist}}^{\text{SSB}} = 20$ ms and for the carrier frequency $f_c = 3.44$ GHz, one can obtain the maximum unambiguous monostatic velocity of ± 1.0901 m/s. For higher velocities, aliasing occurs, leading to unambiguous velocity measurements. This is a main SSB-based sensing limitation in ISAC that can be partially mitigated by $T_{\text{dist}}^{\text{SSB}}$ manipulation. The SSB periodicity can be decreased to the minimum value when a cooperative network is considered. For example, for $T_{\text{dist}}^{\text{SSB}} = 5$ ms and $f_c = 3.44$ GHz, one has the maximum unambiguous monostatic velocity of ± 4.3605 m/s, which can be sufficient in some selected applications. The 5G standard assumes lower frequency utilization, e.g., n28 band (703–748 MHz uplink/758–803 MHz downlink), which allows for wider unambiguous Doppler frequencies to be obtained.

1.3.3 CIR-Based ISAC for FDM Transmission Using 5G Reference Signals

Linear time-variant (LTV) radio channels are characterized by multipath propagation and the Doppler effect [31, 32]. The received signal $r(t)$, see Fig. 1.1, is a sum of $P + 1$ attenuated (by complex value h_p), delayed in time (by τ_p) and shifted in frequency by the Doppler effect (by $f_p^{(D)}$) copies of the transmitted signal $s(t)$ (we assume that the LoS signal is the first, $\tau_0 = 0$, and all delays are relative to it) [31]:

$$r(t) = \sum_{p=0}^{P} h_p s(t - \tau_p) e^{j2\pi f_p^{(D)} t}. \tag{1.18}$$

The channel impulse response (CIR) is usually modeled as [31] follows:

$$h(t, \tau) = \sum_{p=0}^{P} \left[h_p e^{j2\pi f_p^{(D)} t} \right] \delta(\tau - \tau_p), \tag{1.19}$$

where $\delta(t)$ is the Dirac impulse, and t is called a *slow time* while τ—a *fast time*. We see that, due to the Doppler effect, some CIR coefficients h_p can oscillate (be modulated), and the application of some spectral method is required to find the $f_p^{(D)}$ values, e.g., the DFT.

The goal of the communication system is to find the CFR $H(t, f)$, i.e., the Fourier transform of the CIR $h(t, \tau)$ over variable τ, and to equalize its influence upon the

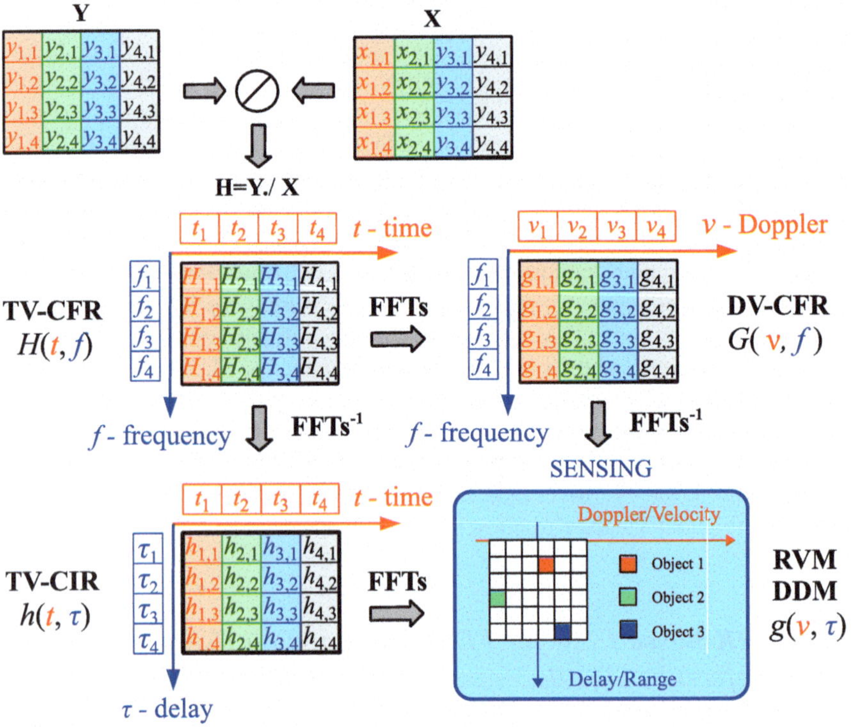

Fig. 1.28 CIR-based OFDM radar utilizing 5G NR signal processing steps: (1) Time-Variant Channel Frequency Response (TV-CFR) calculation using the received signal (Y) and the reference signal (X); (2) Time-Variant Channel Impulse Response (TV-CIR) derivation via inverse fast Fourier transform applied to the columns of the TV-CFR matrix; (3) Range-Velocity Map (RVM) or Doppler-Delay Map (DDM) generation by applying the fast Fourier transform to the rows of the TV-CIR matrix. Alternatively, the RVM/DDM can be obtained via the Doppler-Variant Channel Frequency Response (DV-CFR)

transmitted data (i.e., to *un-do* the channel influence/disturbance)—as already mentioned in the end of Sect. 1.2.1. The task of the radar system is to identify the values of parameters $\{h_p, \tau_p, f_p^{(D)}\}$, and to extract information about the target size/type (from the attenuation h_p), its bistatic range/distance (from the delay τ_p) and bistatic velocity (from frequency shift $f_p^{(D)}$) [25]. In OFDM-based communication systems, the radar task can be easily realized, as explained in Fig. 1.28, which is a continuation/extension of Fig. 1.1. In the telecommunication receiver, first, estimation of the local/momentum TV-CFR (Time-Variant CFR) is calculated for each l-th OFDM symbol based on pilot (p) carriers only: $H_l(\mathbf{k}_p) = Y_l(\mathbf{k}_p) \,./X_l(\mathbf{k}_p)$ (l—OFDM symbol index, $\mathbf{k}$—vector of carrier indexes). Then, the pilot-based CFR estimation is interpolated for all carriers, and $H_l(\mathbf{k}_{all})$ is obtained. Next, the inverse DFT (FFT) of $H_l(\mathbf{k}_{all})$ is computed and $h_l(n)$ results, i.e., estimation of a local TV-CIR (Time-Variant CIR) for the l-th OFDM symbol. Finally, oscillations of each n-th CIR tap, from OFDM symbol to OFDM symbol, must be investigated, e.g., by another DFT, since the Doppler effect causes them. In summary, the range-velocity map $\Psi(R, V)$ is calculated in CIR-based OFDM radars for N_{symb} OFDM symbols with the use of the following two equations (one by one) [33–37]:

$$h_l(n) = \sum_{k=0}^{N_{\text{FFT}}-1} \underbrace{\frac{Y_l(k)}{X_l(k)}}_{H_l(k)} e^{j\frac{2\pi}{N_{\text{FFT}}}kn}, \quad l = 0 \ldots N_{\text{symb}} - 1, \quad n = 0 \ldots N_{\text{FFT}} - 1, \tag{1.20}$$

$$\Psi(R(n), V(m)) = \sum_{l=0}^{N_{\text{symb}}-1} h_l(n) e^{-j\frac{2\pi}{N_{\text{symb}}}lm}, \tag{1.21}$$

where bistatic range $R(n)$ and bistatic velocity $V(m)$ are equal, respectively:

$$R(n) = c\frac{n}{f_s}, \quad n = 0, 1, \ldots, N_{\text{cp}}, \tag{1.22}$$

$$V(m) = m\frac{f_s}{(N_{\text{FFT}} + N_{\text{cp}})N_{\text{symb}}}\frac{c}{f_c}, \quad m = -\frac{N_{\text{symb}}}{2} \ldots \frac{N_{\text{symb}}}{2} - 1. \tag{1.23}$$

In (1.22) and (1.23), the following denotations are used: f_s—sampling rate (in Hertz), f_c—carrier frequency of an OFDM service (in hertz), and c—speed of light (in meters per second). Velocities of objects are calculated only for the first $N_{\text{cp}} + 1$ taps of local CIR estimates since in OFDM systems the N_{cp} value should be bigger than the assumed maximum CIR length. Figure 1.28 also illustrates an alternative approach for computing the range-velocity (RV) map. In this method, FFTs are first applied along the rows of the TV-CFR to produce the Doppler-Variant CFR (DV-CFR), followed by using the IFFTs along its columns to obtain the RV map (also called a delay-Doppler map, DDM).

In Fig. 1.29, the examples, very similar RV maps are presented, calculated for simulated CIR-based (left) and CAF-based (right) 5G NR radars. In the simulation, it was assumed that in the CIR-based radar, all carrier states with data had been correctly reconstructed, and therefore no interpolation was needed during CFR estimation. It should be stressed, however, that the CIR-based approach is mainly suited to the 5G NR with FDD. In TDD, removing the uplink part will cause problems with irregular sampling, which might compromise the sensing capabilities. In turn, the CAF-based approach can be applied to both FDD and TDD 5G versions.

The numerous simulations have shown that application of the 5G NR FDD CIR-based approach is profitable for higher SNRs, while CAF-based radars with long integration time are recommended when stronger noise is present [22, 38]. The results are presented in Figs. 1.30 and 1.31. RV maps were calculated for both CIR and CAF radars, and the PNFR [39] was used for evaluation of the moving object

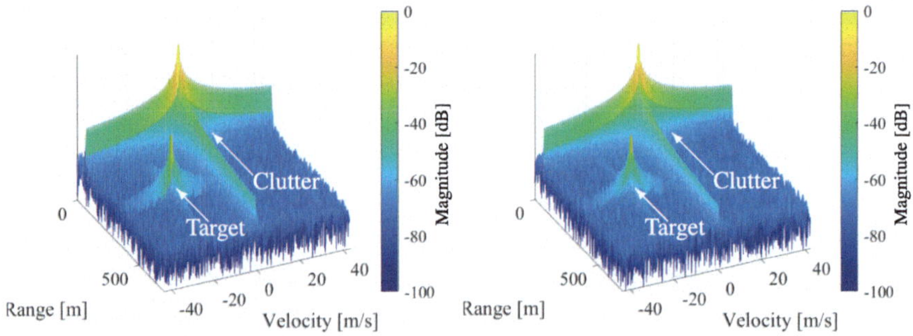

(a) CFR/CIR-based RV map (b) CAF-based RV map

Fig. 1.29 Comparison of radar detection performance for two methods: CFR/CIR based and CAF based. In the simulation, 10 frames of a 5G signal were used, with all OFDM symbols serving as pilots. The channel included two propagation paths: (1) a strong line-of-sight component with zero Doppler, observed as clutter and (2) a weaker target echo with a Doppler shift. Both methods demonstrated identical performance, with comparable noise floor levels and matching clutter and target peak heights

detection quality. The PNFR serves as a measure of detected peak height, defined as the difference between the peak maximum value and the estimated noise floor in dB. In both figures, RoD-K/L denotes the configuration of Radar-on-Demand (RoD) reference signals, sent by the gNB, in which pilots are in every Kth OFDM symbol and every Lth sub-carrier. Parameter **gain** denotes the ratio of the echo peak to the clutter peak. In Fig. 1.30, application of the 5G CSI-RS- and 5G PRS-like pilots is compared with the usage of different RoD pilots and the usage of CAF. The PRS-like signal, designated as a RoD-1/4 pilot, derives its name from its energy spectral properties, which closely resemble those of the 5G PRS. It can be observed that the application of the CAF method is better for lower input SNR, while the application of the CIR method is better for higher input SNR. For the CIR approach, the usage of pilot configurations with higher time–frequency (symbol-carrier) densities is beneficial. This effect is further investigated in Fig. 1.31. Usage of all carriers as pilots results in the highest PNFR.

1.4 Proof of Concept

In this section, the selected measurement examples of 5G-based ISAC are shown.

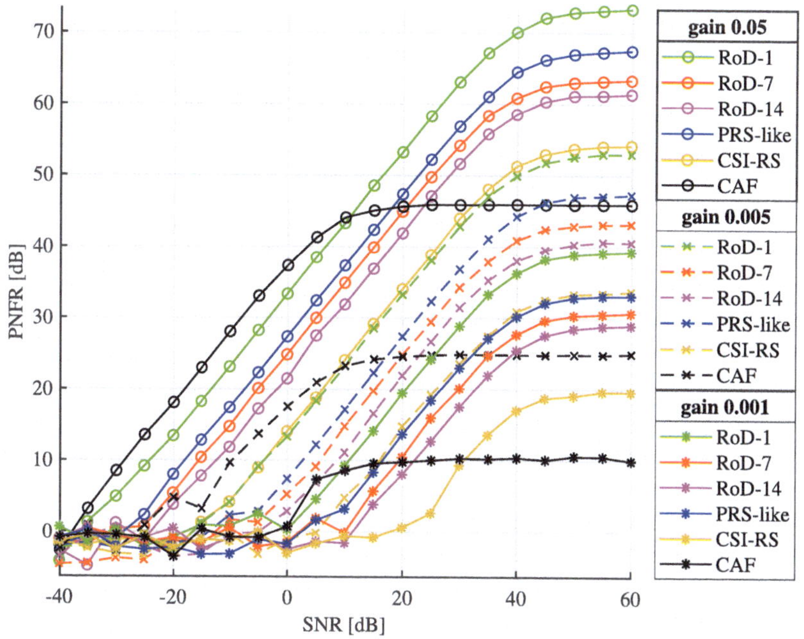

Fig. 1.30 PNFR as function of input SNR for different RoD pilot types, 5G CSI-RS pilots, 5G PRS-like pilots, and for CAF. The simulation was conducted for different target echo gains: 0.05, 0.005, and 0.001, using 10 frames of a 5G signal

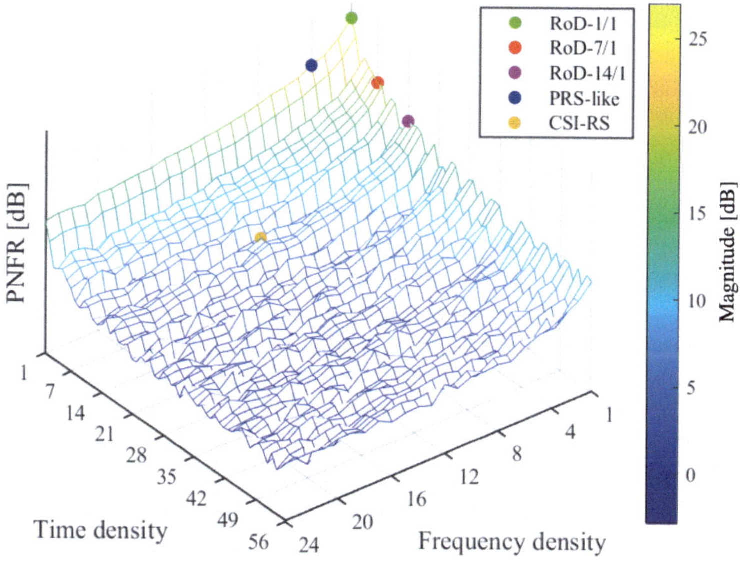

Fig. 1.31 PNFR as function of time and frequency pilot density/repetition. Channel model with target gain equal to 0.008 and SNR = 10 dB used. 10 frames of a 5G waveform

1.4.1 CAF-Based ISAC for High-Bitstream TDD Transmission

The 5G signal is currently being deployed worldwide for both commercial and scientific purposes. The measurements described in this section were conducted within a fully operational 5G testbed network operated by the Łódź University of Technology, Poland. This network comprises five base stations deployed across the university campus and serves as a platform for the DIH5G Digital Innovation Hub project, which focuses on developing novel 5G applications. The system operates in a non-standalone configuration (5G NSA), with the LTE anchor utilizing the 700 MHz frequency band (FDD, 5 MHz downlink and 5 MHz uplink), and NR transceivers operating in the mid-band (3.5 GHz, TDD mode) with a downlink/uplink ratio of 74%/20%. The NR transmission is authorized for a 40 MHz bandwidth within the 3420–3460 MHz frequency range, with each base station featuring a nominal output power of 50 W.

The testbed infrastructure is built entirely from commercially available components, including the 5G core, fiber-optic backbone, network management system, radio access units, and antennas, all under the full administrative control of the university. The NR base stations are equipped with active Massive MIMO AIR 6488 antennas, supporting spatial beamforming and multiplexing across 64 transmitter/receiver branches. These antennas offer a traffic beam gain of up to 23 dBi.

Several experiments were conducted to verify the possibility of target detection. In the next part of this chapter, we focus on two selected examples, namely, car and drone detection.

The two-channel passive radar demonstrator built using the Ettus USRP X310 software-defined radio platform and synchronized using a GPS time reference was used. Two Commercial Off-The-Shelf (COTS) log-periodic antennas handled signal reception, both horizontally polarized (the polarization diversity was neglected in the experiment).

Each signal path included low-noise amplification and band-pass filtering to ensure proper signal dynamics and minimize the noise floor. The reference channel employed a single low-noise amplifier with a gain of approximately 10 dB, whereas the surveillance channel used an amplifier with twice the gain. This configuration prevented saturation in the reference path while maintaining a sufficient signal level for the surveillance channel to ensure reliable post-processing. Figure 1.32 [1] shows a simplified schematic of the recording setup. This signal acquisition setup was identical in the two analyzed cases; the only difference was the cooperative target. Figure 1.33 delineates the range-velocity maps for car (a) and drone (b).

The presented experiments confirm the feasibility of detecting moving targets using 5G signals in a passive radar configuration. Both cooperative objects—a car and a drone—were successfully detected within the operational 5G testbed network. The experiment also demonstrated the possibility of the 5G base stations as illuminators of opportunity. However, it was observed that to ensure a sufficiently strong signal

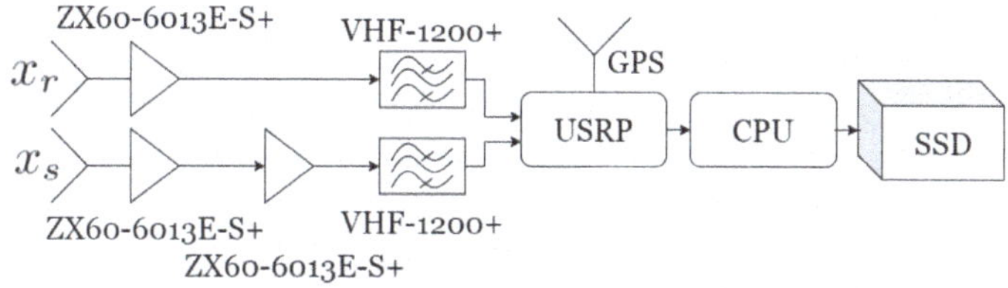

Fig. 1.32 A block diagram of the passive radar demonstrator used in the experiment. The system was built of commercial-off-the-shelf components. Reproduced from [1]

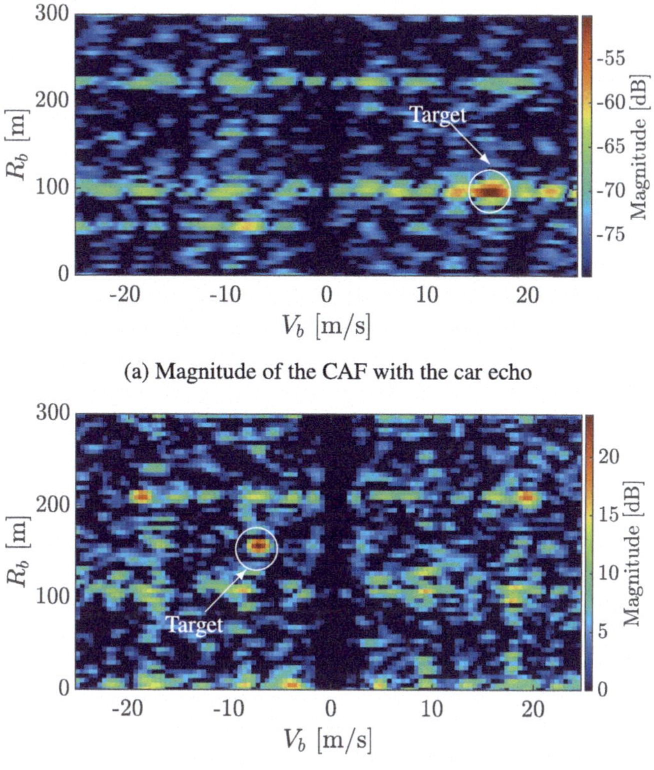

(a) Magnitude of the CAF with the car echo

(b) Magnitude of the CAF with the drone echo.

Fig. 1.33 Real-life examples of the target detection using the real-life 5G network as an illuminator. In two cases, the target was successfully detected by the passive radar system

for target detection, forced data transmission must be maintained in the network to keep the downlink active.

Additionally, the experiments highlight the importance of polarization alignment between the radar receiver and the base station antennas, as polarization mismatch can degrade detection performance. Despite neglecting polarization diversity in this setup, the detection results successfully validated the overall approach. These findings confirm that the radar system architecture and the data transmission environment operated reliably, enabling effective joint communication and sensing.

1.4.2 CAF-Based ISAC for Low-Bitstream TDD&FDD Transmission Using SSB

To investigate the feasibility of joint communication and sensing using SDR, an experimental 5G ISAC transceiver node was implemented based on programmable radio hardware. In this setup, the transmitter and receiver functionalities, including active radar sensing, were realized using a USRP X310 device equipped with two UBX-160 daughterboards. The SDR was controlled by a host PC running Windows 10, powered by an Intel Core i7-9700 CPU (2.66 GHz), with 32 GB RAM and SSD storage.

LabVIEW software was used to coordinate simultaneous signal transmission and data acquisition. A pre-generated 5G downlink waveform was transmitted from the USRP's TX channel, amplified using a high-power amplifier, and radiated via a GigaEter 19 Econo WideBand antenna with a beamwidth of 18° in both azimuth and elevation. The transmitted signal was also coupled into one of the USRP's RX channels to enable radar functionality and stored as a reference signal. A second, identical antenna was used to receive signals reflected from objects within the illuminated area. After passing through a low-noise amplifier and band-pass filter, the echo signal was fed into the second RX channel and saved to disk for offline sensing analysis.

This configuration emulates a 5G base station enhanced with sensing capabilities, forming the basis for further studies on ISAC signal processing and detection performance.

An emulated UE node was implemented in a similar fashion. It featured two receiving channels: one for capturing the direct signal from the 5G ISAC transceiver and another with its antenna oriented toward the surveillance area. Signals from both channels were coherently received and stored on the PC's hard drive. The node was designed to receive and decode communication data from the emulated base station while incorporating additional passive radar sensing functionality.

During the experiment, 5G-compliant transmission was flawlessly conducted between the two nodes. The signal, which carried information bits, also functioned as illumination for both active and passive sensing. Both nodes simultaneously detected a vehicle driving on a nearby road. Integrating the detection data improves the preci-

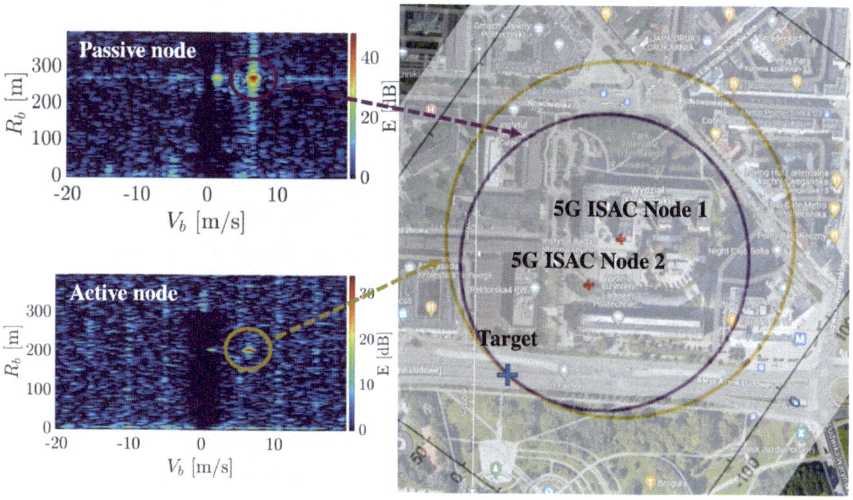

Fig. 1.34 Results of the detection data fusion from both active and passive 5G ISAC nodes. Thanks to the specific geometry involving both the active and passive nodes, the target position was precisely estimated. Reproduced from [10]

sion of target localization, as illustrated in Fig. 1.34. In the figure, the purple ellipse represents the target's possible location as detected by the passive node, the dark yellow circle indicates the position determined by the active node, and their intersection identifies the exact target position.

The experiment demonstrates how radar functionality can be seamlessly integrated into the existing 5G standard. This dual-use system offers significant advantages by utilizing the same signal for communication and sensing. It saves energy, eliminates the need for redundant equipment, and addresses the issue of crowded spectrum. Establishing a network of sensors, such as one active node and one or more passive nodes, communicating via the 5G standard creates an efficient system capable of precise target localization.

1.4.3 CIR-Based ISAC for FDM Transmission Using 5G Reference Signals

The same SDR setup described in Sect. 1.4.2 (consisting of the USRP X310 transmitter and receiver connected to a PC) was employed to validate the concept of a CIR-based passive radar utilizing 5G NR gNB signals, as detailed in [38]. The antenna placement remained unchanged, with one antenna mounted on the roof of a university building and the other located in the university parking area. The pre-recorded 5G waveform included both CSI-RS and RoD pilot symbols. The primary objective of these measurements was to detect vehicles moving on a high-speed road

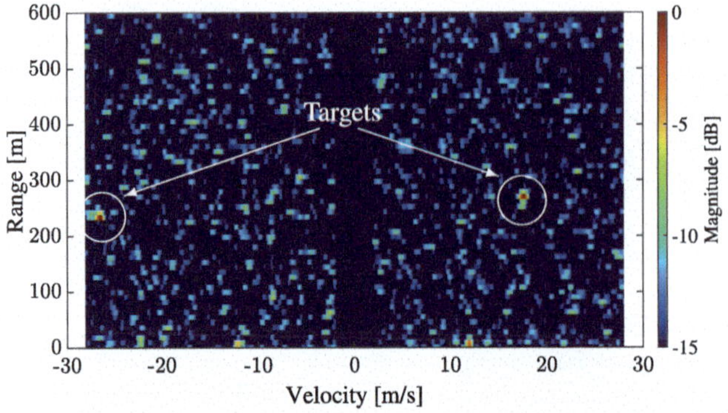

(a) RV map for the CIR-based processing with ECA-CD clutter removal

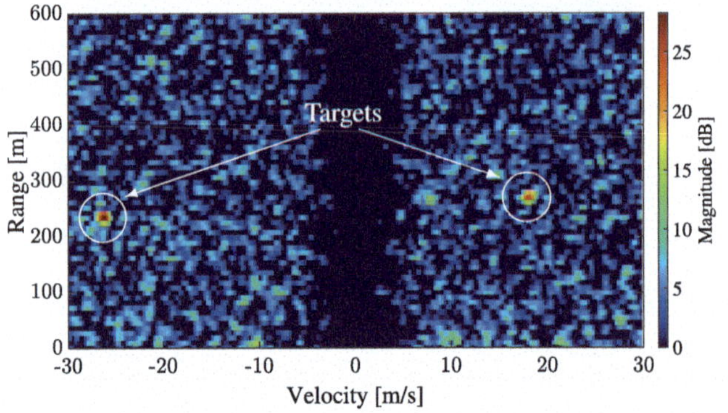

(b) RV map for the CAF-based processing with clutter removal via the lattice filter

Fig. 1.35 RV maps generated using two approaches: **a** CIR-based processing with ECA-CD clutter removal and **b** CAF-based processing with clutter removal via the lattice filter. Both methods successfully detect two targets, indicated by white circles. Reproduced from [38]

located near the receiver. On the left side of Fig. 1.35, the RV map obtained using the CIR/CFR-based radar is presented, clearly showing two detected targets. To suppress clutter, the Extensive Cancellation Algorithm by Carrier and Doppler (ECA-CD) [39, 40] was applied. To verify the accuracy of these results, a second RV map was generated using the previously described CAF-based method in combination with the lattice filter for clutter suppression, which confirmed the detection of the same two targets.

1.5 Conclusions

In this chapter, the concept and experimental evaluation of energy-efficient 5G-based ISAC systems have been presented. Although the 5G standard is extensive, it also offers considerable flexibility, allowing configurations that enable simultaneous communication and sensing. However, as the standard is primarily designed for communication, certain limitations must be addressed to support effective radar functionality. These include the use of time- or frequency-division duplexing (the former introduces gaps in the downlink signal, while the latter limits signal bandwidth), content-dependent transmission (which can result in insufficient illumination when no data is being transferred), and base station beamforming. These characteristics complicate the straightforward implementation of sensing capabilities and require additional considerations to enable accurate and efficient target localization during data transmission.

Nonetheless, 5G networks offer high-resolution sensing and high-throughput data transmission compared to other commercial technologies, making the effort to adapt them for joint communication and sensing worthwhile.

This chapter demonstrated how existing 5G signals can be leveraged for situational awareness in two scenarios. The first explored both passive and active sensing, utilizing the unmodified 5G signal without altering the physical layer. While this approach necessitates a tailored signal processing chain for sensing, the communication subsystem remains unaffected. In contrast, when the signal is slightly modified to enhance sensing performance, it may require minor adjustments to the communication component. Still, both functionalities can generally coexist without mutual interference.

In light of these findings, the overall approach aligns closely with the goals of the Sustainable Development Agenda, particularly in areas where reliable data collection and connectivity are critical. By enabling efficient environmental monitoring, real-time data acquisition, and low-power communication, integrated systems like ISAC support resilient urban infrastructure and enhanced situational awareness. These capabilities are especially valuable in resource-constrained settings, where infrastructure must be both cost-effective and energy-efficient.

Acknowledgements The authors would like to thank all those who contributed and assisted with measurements and data processing. Special thanks go to Prof. Piotr Korbel, Prof. Sławomir Hausman, Prof. Mateusz Malanowski, Marek Płotka, Piotr Tomikowski, and Witold Duda.

References

1. Samczyński, P., Abratkiewicz, K., Płotka, M., Zieliński, T.P., Wszołek, J., Hausman, S., Korbel, P., Księżyk, A.: 5G network-based passive radar. IEEE Trans. Geosci. Remote Sens. **60**, 1–9 (2022)

2. Księżyk, A., Płotka, M., Abratkiewicz, K., Maksymiuk, R., Wszołek, J., Samczyński, P., Zieliński, T.P.: Opportunities and limitations in radar sensing based on 5G broadband cellular networks. IEEE Aerosp. Electron. Syst. Mag. **38**(9), 4–21 (2023)
3. Liu, B., Lyu, X., Fan, W.: Analysis of 5G signal for radar application. In: Journal of Physics: Conference Series, vol. 2356, p. 012027. IOP Publishing (2022)
4. Ilioudis, C.V., Clemente, C.: 5G based passive radar: a feasibility study. IET Conf. Proc. **208–213**, 2023 (2022)
5. Kanhere, O., Goyal, S., Beluri, M., Rappaport, T.S.: Target localization using bistatic and multistatic radar with 5G NR waveform. In: 2021 IEEE 93rd Vehicular Technology Conference (VTC2021-Spring), pp. 1–7 (2021)
6. Kiviranta, M., Moilanen, I., Roivainen, J.: 5G radar: scenarios, numerology and simulations. In: 2019 International Conference on Military Communications and Information Systems (ICMCIS), pp. 1–6. IEEE (2019)
7. Thoma, R.S., Andrich, C., Del Galdo, G., Dobereiner, M., Hein, M.A., Kaske, M., Schafer, G., Schieler, S., Schneider, C., Schwind, A., et al.: Cooperative passive coherent location: a promising 5G service to support road safety. IEEE Commun. Mag. **57**(9), 86–92 (2019)
8. Baquero Barneto, C., Riihonen, T., Turunen, M., Anttila, L., Fleischer, M., Stadius, K., Ryynänen, J., Valkama, M.: Full-duplex OFDM radar with LTE and 5G NR waveforms: challenges, solutions, and measurements. IEEE Trans. Microw. Theory Tech. **67**(10), 4042–4054 (2019)
9. Dwivedi, M., Hulede, I.E.L., Venegas, O., Ashdown, J., Mukherjee, A.: 5G-based passive radar sensing for human activity recognition using deep learning. In: 2024 IEEE Radar Conference (RadarConf24), pp. 1–6 (2024)
10. Maksymiuk, R.: Emerging trends in radar: 5G/6G joint communication and active & passive sensing. IEEE Aerosp. Electron. Syst. Mag. 1–4 (2025)
11. Fang, X., Feng, W., Chen, Y., Ge, N., Zhang, Y.: Joint communication and sensing toward 6G: models and potential of using MIMO. IEEE Internet Things J. **10**(5), 4093–4116 (2023)
12. Zielinski, T.P.: Starting Digital Signal Processing in Telecommunication Engineering. A Laboratory-Based Course, Springer Nature (2022)
13. ETSI. TS 38.201: 5G NR; Physical layer; General description. standard Version 17.0.0, ETSI (2022)
14. ETSI. TS 38.202: 5G NR; Services provided by the physical layer. standard Version 17.2.0, ETSI (2022)
15. ETSI. 5G NR Physical channels and modulation. standard Version 16.2.0, ETSI (2020)
16. ETSI. TS 38.104: 5G NR; Multiplexing and channel coding. standard Version 16.2.0, ETSI (2020)
17. ETSI. TS 38.213: 5G NR; Physical layer procedures for control. standard Version 17.2.0, ETSI (2022)
18. ETSI. TS 38.214: 5G NR; Physical layer procedures for data. standard Version 17.1.0, ETSI (2022)
19. ETSI. TS 38.215: 5G NR; Physical layer measurements. standard Version 18.2.0, ETSI (2024)
20. ETSI. TS 38.104: 5G NR; Base Station (BS) radio transmission and reception. standard Version 16.7.0, ETSI (2021)
21. Abratkiewicz, K., Księżyk, A., Płotka, M., Samczyński, P., Wszołek, J., Zieliński, T.P.: SSB-based signal processing for passive radar using a 5G network. IEEE J. Sel. Top. Appl. Earth Obs. Remote. Sens. **16**, 3469–3484 (2023)
22. Wypich, M., Zielinski, T.P.: 5G NR based radar-on-demand using channel impulse response estimate. In: 2023 Signal Processing Symposium (SPSympo), pp. 220–225 (2023)
23. Maksymiuk, R., Samczyński, P., Abratkiewicz, K.: Impact of uplink traffic on 5G-based passive radar. In: 2023 Signal Processing Symposium (SPSympo), pp. 117–122 (2023)
24. Griffiths, H.D., Baker, C.J.: An Introduction to Passive Radar, Artech House radar library. Artech House Publishers (2017)
25. Malanowski, M.: Signal Processing for Passive Bistatic Radar. Artech House radar library. Artech House (2019)
26. Flandrin, P.: Explorations in Time-Frequency Analysis. Cambridge University Press (2018)

27. Ergen, M.: Mobile Broadband - Including WiMAX and LTE, 1st edn. Springer Publishing Company, Incorporated (2009)
28. Baraniuk, R.G., Flandrin, P., Janssen, A.J.E.M., Michel, O.J.J.: Measuring time-frequency information content using the Renyi entropies. IEEE Trans. Inf. Theory **47**(4), 1391–1409 (2001)
29. Maksymiuk, R., Abratkiewicz, K., Samczyński, P., Płotka, M.: Rényi entropy-based adaptive integration method for 5G-based passive radar drone detection. Remote. Sens. **14**(23) (2022)
30. Zaidi, A., Athley, F., Medbo, J., Gstavsson, U., Durisi, G., Chen, X.: 5G PHYSICAL LAYER: Principles, Models and Technology Components. ACADEMIC PR INC (2018)
31. Hlawatsch, A., Matz, G. (Eds.).: Wireless Communications over Rapidly Time-Varying Channels. Academic Press (2011)
32. Matz, G., Bölcskei, H., Hlawatsch, F.: Time-frequency foundations of communications. IEEE Signal Process. Mag. 87–96 (2013)
33. Sturm, C., Pancera, C., Zwick, T., Wiesbeck, W.: A novel approach to OFDM radar processing. In: IEEE Radar Conference (2009)
34. Braun, M., Sturm, Ch., Niethammer, A., Jondral, F.K.: Parametrization of joint OFDM-based radar and communication systems for vehicular applications. In: IEEE 20th International Symposium on Personal, Indoor and Mobile Radio Communications (2009)
35. Sturm, C., Pancera, E., Zwick, T., Wiesbeck, W.: Waveform design and signal processing aspects for fusion of wireless communications and radar sensing. Proc. IEEE **99**(7), 1236–1259 (2011)
36. Hakobyan, G.: Orthogonal Frequency Division Multiplexing Multiple-Input Multiple-Output Automotive Radar with Novel Signal Processing Algorithms. PhD thesis, Universität Stuttgart (2018)
37. Hakobyan, G., Yang, B.: High-performance automotive radar. IEEE Signal Process. Mag. **36**, 32–44 (2019)
38. Wypich, M., Maksymiuk, R., Zielinski, T.P.: 5G-based passive radar utilizing channel response estimated via reference signals. IEEE Trans. Radar Syst. **3**, 511–519 (2025)
39. Colone, F., O'Hagan, D.W., Lombardo, P., Baker, C.J.: A multistage processing algorithm for disturbance removal and target detection in passive bistatic radar. IEEE Trans. Aerosp. Electron. Syst. **45**(2), 698–722 (2009)
40. Schwark, C., Cristallini, D.: Advanced multipath clutter cancellation in OFDM-based passive radar systems. In: 2016 IEEE Radar Conference (RadarConf), pp. 1–4 (2016)

Chapter 2
Integrated ISAC-RIS Systems for Next-Generation Connectivity Enhancement

Manojkumar B. Kokare, Neha Sharma, Sumit Gautam, and R. Swaminathan

Abstract This chapter provides an extensive survey of recent advancements in integrated sensing and communication (ISAC) systems facilitated by reconfigurable intelligent surfaces (RIS), emphasizing their transformative potential for next-generation wireless networks. A systematic overview is presented, covering the fundamental principles, key enabling technologies, and possible state-of-the-art implementations of ISAC-RIS architectures. The dual functionality of ISAC-RIS in enhancing both sensing accuracy and communication reliability through intelligent signal manipulation and environment-aware beamforming is highlighted. A detailed taxonomy is introduced, categorizing existing works based on application domains. In this chapter, we provide a critical evaluation of different system architectures utilizing ISAC-RIS for connectivity enhancement across various use cases, including unmanned aerial vehicle (UAV) systems, vehicle-to-everything (V2X) communications, Internet-of-things (IoT) applications, and secure transmission. It also presents a simplified case study of ISAC-RIS systems, focusing on performance analysis through outage probability (OP) and exploring optimization techniques aimed at maximizing the overall sum rate. In addition, emerging trends are discussed, including the integration of simultaneous transmitting and reflecting (STAR)-RIS systems, beyond diagonal RIS, simultaneous wireless information and power transfer (SWIPT) systems for energy efficiency enhancement and aspects of machine learning (ML) for adaptive RIS configuration to improve overall system performance. The chapter concludes with a review of open research challenges and future directions,

M. B. Kokare and N. Sharma: These authors contributed equally to this work.

M. B. Kokare · N. Sharma · S. Gautam (✉) · R. Swaminathan
Department of Electrical Engineering, Indian Institute of Technology Indore, Indore, Madhya Pradesh, India
e-mail: sumit.gautam@iiti.ac.in

M. B. Kokare
e-mail: phd2201102016@iiti.ac.in

N. Sharma
e-mail: phd2201102021@iiti.ac.in

R. Swaminathan
e-mail: swamiramabadran@iiti.ac.in

A. K. Mishra and V. Bhatia (eds.), *Integrated Sensing and Communication for Sustainable Development Goals*, Signals and Communication Technology,
https://doi.org/10.1007/978-981-92-0209-6_2

providing valuable insights for researchers at the intersection of sensing, communication, and smart radio environments.

Keywords Integrated sensing and communication (ISAC) · Internet of things (IoT) · Outage probability (OP) · Reconfigurable intelligent surfaces (RIS) · Unmanned aerial vehicles (UAV) · Vehicle-to-everything (V2X)

2.1 Introduction

Next-generation wireless networks, encompassing beyond 5G (B5G) and sixth-generation (6G) systems, are envisioned as pivotal enablers for a broad spectrum of emerging and future-driven applications. These applications impose stringent requirements not only on the quality and reliability of wireless connectivity but also on the accuracy, robustness, and adaptability of environmental sensing capabilities. Among the many visionary aspects associated with B5G and 6G networks, a consistent and prominent theme is the significantly enhanced role of sensing, a function expected to evolve into a core component of the future wireless ecosystem. Although the exploration of future wireless systems remains at a relatively early, speculative stage, current technological trends clearly indicate a strong readiness to embrace integrated sensing functionalities within B5G and 6G frameworks. Specifically, both radio sensing and communication (S&C) systems are converging toward the use of higher frequency bands, deployment of larger and more sophisticated antenna arrays, and advancements in miniaturized hardware. This convergence is progressively aligning the two domains in terms of hardware architectures, channel propagation characteristics, and signal processing (SP) methodologies [1].

This alignment creates a compelling opportunity to implement sensing functionalities using existing and evolving wireless communication infrastructures. Future wireless networks are anticipated to transcend their traditional role of data transmission and provide ubiquitous environmental sensing services, enabling capabilities such as real-time monitoring, object detection, and even high-resolution imaging of the surrounding environment. The ability to sense and interpret environmental data will play a crucial role in building intelligence into future smart environments, thereby supporting a multitude of location-aware and context-aware applications. These include, but are not limited to, vehicle-to-everything (V2X) communications, smart homes, intelligent manufacturing systems, remote sensing, environmental surveillance, and human–computer interaction.

In pursuit of these advanced capabilities, there is a critical need to jointly design S&C operations within B5G and 6G networks. This necessity has led to the emergence of integrated sensing and communications (ISAC) as a prominent research paradigm. ISAC has two fundamentally different objectives, S&C: sensing aims to infer information from noisy observations of the physical environment, whereas communication is concerned with the reliable transfer of information through designed waveforms over noisy channels. ISAC seeks to unify these traditionally separate functions into a cohesive system, thereby enabling mutual performance enhancements and resource optimization [2].

From a system design perspective, ISAC is expected to significantly enhance spectral and energy efficiencies by facilitating the reuse of hardware, bandwidth, and signaling resources for dual purposes. By consolidating S&C into a singular platform, ISAC reduces the need for dedicated subsystems that historically competed for limited resources. More importantly, ISAC advocates for a deep integration paradigm, wherein the two functionalities are no longer treated as distinct end-goals but are co-optimized to support one another. This includes communication-assisted sensing, where transmitted signals serve dual roles in information transfer and environment probing, and sensing-assisted communication, where environmental insights can improve link reliability and system performance.

Despite its recent surge in interest within both academia and industry, the foundational concepts underlying ISAC have been explored under various terminologies over the past decades. These include radar-communications (RadCom), joint communication and radar (JCR), joint radar and communication (JRC), and dual-functional radar-communications (DFRC). While these terms may differ in emphasis and scope, they primarily focus on the integration of radar-based sensing with communication systems. For consistency and generality, this overview adopts the term ISAC as a unified framework encompassing all joint design approaches that integrate radar sensing with wireless communication functionalities [3].

To address the inherent challenges faced by ISAC systems, particularly under adverse radio propagation conditions, recent research advocates the integration of reconfigurable intelligent surfaces (RISs) into ISAC frameworks. RISs, composed of engineered electromagnetic meta-materials, possess the unique capability to manipulate incident electromagnetic waves by imposing precise phase shifts and controlled delays on the reflected signals [4]. This programmable control enables RISs to enhance the signal-to-noise ratio (SNR) at the receiver, thereby improving both communication reliability and sensing accuracy. Consequently, RIS-assisted ISAC systems offer a promising solution for overcoming signal degradation in complex environments, paving the way for more efficient and robust joint S&C performance.

2.1.1 Fundamentals of Radar and Communications

In both radar and communication (R&C) systems, the fundamental architecture can be generally delineated into three primary components: the transmitter (Tx), the propagation channel, and the receiver (Rx). The Tx is responsible for generating electromagnetic (EM) waves, which subsequently propagate through the physical medium referred to as the channel before being intercepted by the Rx. The received EM signal typically undergoes various forms of distortion due to the channel's inherent characteristics, such as fading, scattering, and multipath effects [5].

In conventional communication systems, the Tx and Rx are typically spatially separated to facilitate bidirectional or unidirectional information exchange. In contrast, radar systems exhibit more flexibility in their configuration. Specifically, radar Txs and Rxs may either share a common physical location, referred to as a monostatic radar configuration, or be situated at distinct spatial positions, leading to what is known as a bistatic radar setup. Furthermore, more sophisticated system architectures may incorporate multiple Txs and Rxs. This results in multi-user communication networks or multistatic radar systems, depending on the specific application.

To facilitate signal analysis and processing in both R&C contexts, EM waves are often modeled via their electric field representation. This is typically expressed as a complex-valued function of time, which captures both the amplitude and phase information of the signal. Within this unified framework, the principal objectives of R&C systems can be formally characterized as follows:

- **Information Extraction in Radar Systems**: The primary goal is to retrieve meaningful information regarding one or more targets based on the received signal, with the assumption that the characteristics of the transmitted waveform are known a priori. This includes estimating parameters such as target range, velocity, and angle of arrival.
- **Information Transmission in Communication Systems**: The focus is on accurately recovering the data embedded in the transmitted signal at the Rx end. This process typically assumes knowledge of the channel response, which enables the mitigation of channel-induced distortions and facilitates reliable decoding of the transmitted message.

Why Integrate Sensing and Communication?

The integration of S&C functionalities into a unified framework commonly referred to as ISAC is witnessing accelerated momentum in both academic research and industrial development. This growing interest is driven by the increasing demand for intelligent, resource-efficient, and context-aware wireless systems, especially in the era of 6G and beyond.

Traditionally, sensing (e.g., radar, localization, and environment mapping) and communication (e.g., data exchange and network connectivity) are designed and operated as separate systems. However, this disjoint approach leads to inefficient utilization of key resources such as spectrum, hardware, and energy. ISAC offers a paradigm shift by jointly designing and implementing both functionalities, enabling them to share common physical and spectral resources and even enhance each other's performance.

2.1.2 What is ISAC?

ISAC represents a transformative design paradigm in which radio-frequency (RF) sensing and wireless communication functionalities are jointly developed and operated within a unified system architecture. This convergence enables efficient utilization of increasingly congested spectral and hardware resources, while also allowing for potential mutual performance enhancements between S&C processes. In addition to defining the concept itself, ISAC also encompasses the suite of enabling technologies that facilitate such integration.

The term “integration” in the context of ISAC broadly refers to any functional or structural coupling between S&C systems, whether partial or complete. For instance, in a distributed wireless sensor network, integration may manifest through shared hardware between S&C modules at the node level. Alternatively, in secondary surveillance radar systems, integration can occur at the signaling level, as the radar system communicates with airborne transponders. Another example includes cognitive radar systems operating within communication frequency bands, which represent cases of spectrum-level integration [6].

While ISAC offers significant potential to enhance efficiency across multiple dimensions, including spectral, temporal, energy, hardware, and signaling domains, the actual benefits depend on the specific layer or aspect at which the integration is realized. The core principle underpinning ISAC is that a single radio signal can serve a dual purpose: transmitting data to a communication Rx while simultaneously extracting information about the surrounding environment from the reflected or backscattered signal. The most advanced form of ISAC is achieved when both S&C functions are embedded within a unified waveform, enabling simultaneous operation with minimal overhead. This tightly coupled configuration yields comprehensive improvements in resource utilization, such as

1. **Spectral Efficiency**: Through shared frequency bands for both functions.
2. **Hardware Efficiency**: By using common RF chains, antennas, and processing units.
3. **Energy Efficiency**: Via reduced power consumption through consolidated operations.
4. **Latency and Signaling Overhead**: By minimizing redundant transmissions and control messaging.

The organization of this chapter is as follows. This section, i.e., Sect. 2.1 introduced the fundamental concepts and key characteristics of ISAC technology. Section 2.2 outlines the various use cases of ISAC. The challenges faced by ISAC systems are discussed in Sect. 2.3. Section 2.4 explores the integration of RIS with ISAC systems. Section 2.5 highlights specific use cases of ISAC-RIS integration. The system model of RIS-assisted ISAC is detailed in Sect. 2.6, along with its performance analysis and optimization framework. Section 2.7 addresses the key challenges associated with ISAC-RIS systems. Section 2.8 discusses the future prospects of RIS in ISAC-enabled 6G networks. Finally, Sect. 2.9 concludes the chapter.

2.2 ISAC Use Cases

In this section, we illustrate our ISAC vision for future wireless networks. In particular, we consider six potential ISAC application scenarios, discussed as follows:

1. **Civilian Applications**: Accurate localization has been integral to cellular networks from 1G to the envisioned 6G; however, current systems such as 4G and 5G are constrained by limited bandwidth and antenna configurations, resulting in meter-level accuracy that primarily complements global navigation satellite systems (GNSS), e.g., global positioning system (GPS) and navigation with Indian constellation (NavIC) [7]. With the growing demands of autonomous driving, augmented reality, and smart city infrastructures, integrated sensing and communication (ISAC) can significantly enhance localization precision by leveraging wider bandwidths, millimeter-wave (mmWave) frequencies, and massive multiple-input multiple-output (mMIMO) arrays. Moreover, radio-frequency (RF) imaging offers high-resolution, privacy-preserving, and weather-resilient environmental sensing capabilities, enabling advanced surveillance and mapping with unprecedented accuracy. The proliferation of unmanned aerial vehicles (UAVs) further underscores the need for robust detection and management frameworks, as their compact size, low-altitude operations, and non-cooperative behavior challenge conventional radar systems. In this regard, ISAC-enabled technologies promise enhanced UAV detection, identification, and tracking, thereby addressing critical civilian and security concerns.
2. **Intelligent Living Environments**: In modern indoor environments such as homes, vehicles, and smart buildings, there is a growing demand for intelligent and adaptive systems that ensure comfort, safety, and convenience. To this end, smart IoT devices are increasingly being designed to perceive and interpret occupants' presence and physiological states in a non-intrusive manner. Leveraging the ubiquity and privacy-preserving features of wireless standards including Wi-Fi, LoRa-WAN, and 5G-NR, indoor sensing has emerged as a scalable solution by re-purposing communication signals for detecting human presence, gestures, motion, and even vital signs without the need for cameras or wearables [8]. The integration of sensing and communication (S&C) into an ISAC framework enables IoT systems to operate as ambient intelligence agents, supporting applications such as activity recognition, in-home healthcare, intrusion detection, and driver attention monitoring. Several of these functions are already embedded in consumer-grade products, and as ISAC technologies evolve, their ability to merge communication infrastructures with real-time environmental awareness will be pivotal to realizing context-aware and intelligent living environments.
3. **Next-Gen Industrial Automation**: The integration of wireless networking technologies into industrial environments such as construction sites, automotive plants, and automated production lines has accelerated the growth of the industrial IoT, driving substantial gains in automation, efficiency, and precision. These scenarios involve dense connectivity among network nodes and autonomous robots, requiring ultra-reliable and low-latency communication (URLLC) to ensure real-time responsiveness. Within this context, ISAC offers distinct advantages by embedding sensing into communication systems, thereby enabling robotic agents and factory nodes to navigate, coordinate, and map complex environments while reducing reliance on separate sensing infrastructures and minimizing signaling overhead [9]. Representative ISAC-enabled functions include

swarm navigation, AGV platooning, and high-resolution environmental imaging, all of which must meet the stringent requirements of Industry 4.0, namely, ultra-low latency, ultra-high reliability, and massive machine-type connectivity. By fulfilling these constraints, ISAC facilitates seamless orchestration of smart factories, resulting in safer, more flexible, and highly productive industrial automation systems.

4. **Connected Vehicle Ecosystems**: Autonomous vehicles (AVs) are expected to transform transportation by enhancing road safety, traffic efficiency, energy consumption, and environmental sustainability. These systems rely on integrated communication transceivers and environmental sensors that enable real-time perception and vehicle-to-everything (V2X) interactions with roadside units (RSUs), nearby vehicles, and pedestrians. The integration of S&C functions within the ISAC paradigm offers a unified framework that reduces hardware complexity, spectrum congestion, and electromagnetic compatibility issues while improving scalability, reliability, and energy efficiency [10]. ISAC-enabled V2X communication supports rapid platooning, secure low-latency data exchange, and advanced capabilities such as simultaneous localization and mapping, which are essential for cooperative and coordinated driving. Furthermore, RSU-based ISAC infrastructures extend sensing beyond a vehicle's line of sight by sharing real-time environmental data, thereby enhancing situational awareness and enabling proactive responses to potential hazards. Collectively, ISAC-enabled vehicular networks represent a key enabler for intelligent transportation systems, driving safer, more efficient, and sustainable mobility solutions [11].
5. **Smart Environmental Surveillance**: The propagation characteristics of wireless signals inherently carry rich environmental information, enabling their use for real-time monitoring of atmospheric conditions such as humidity, rainfall, and particulate concentrations. Higher-frequency bands, particularly mmWave, exhibit greater sensitivity to environmental changes due to their proximity to water vapor absorption bands, allowing variations in path-loss patterns across dense urban cellular networks to be correlated with meteorological and ecological factors including rainfall intensity, air pollutants, and even insect swarms [12]. By enhancing cellular networks with integrated sensing, existing wireless infrastructures can be transformed into large-scale, low-cost, and continuously operating environmental observation systems, offering broad spatial coverage and high temporal resolution. Unlike traditional weather stations that require dedicated sensors, ISAC-enabled networks provide environmental insights as a byproduct of communication operations. As wireless systems evolve toward THz and sub-THz frequencies in 6G and beyond, their heightened sensitivity to atmospheric and biological phenomena will further extend the scope of urban ecological monitoring, aligning with the vision of sustainable, intelligent cities.
6. **Advanced Sensing for Geoscientific Applications**: Radar systems mounted on satellites and airborne platforms have long been central to geoscientific applications, offering high-resolution, all-weather, day-and-night imaging that supports environmental monitoring, Earth system science, change detection, 4D mapping, security surveillance, and planetary exploration [13]. Most operational platforms

employ synthetic aperture radar (SAR) with waveforms such as linear frequency modulated chirps or OFDM, and recent advances in waveform design now allow embedding communication data within radar signals, enabling low-rate broadcasting or covert communication alongside sensing. In disaster response, UAVs equipped with radars, thermal sensors, and communication payloads are vital for damage assessment, search-and-rescue, and emergency connectivity, though such multi-payload requirements often constrain endurance and range. ISAC provides an efficient solution by integrating S&C. functions into a unified system, thereby improving hardware and energy efficiency and extending UAV operation. Furthermore, cooperative ISAC among multiple UAVs or satellites enables swarm-based SAR with distributed apertures, supporting high-resolution imaging and enhanced situational awareness through coordinated sensing and communication.

2.3 Problems Faced by ISAC

Although ISAC has attracted significant research attention in recent years and has been extensively investigated from multiple perspectives, numerous open challenges remain unaddressed. Despite the notable progress, several critical issues particularly related to fundamental trade-offs, SP methodologies, and network-level integration continue to pose substantial barriers to the full realization of ISAC systems. These unresolved problems highlight the need for sustained and rigorous research efforts to unlock the full potential of ISAC in next-generation wireless networks [14].

In what follows, we provide an overview of key open research challenges across the aforementioned domains. These challenges not only demand innovative theoretical frameworks but also call for practical design methodologies that can bridge the gap between S&C in realistic deployment scenarios.

1. **ISAC Rx Design**: Designing effective receiver architectures for ISAC systems is challenging in rich multipath and dense scattering environments, such as indoors. A key difficulty lies in distinguishing between sensing echoes and communication signals, which are superimposed in the received waveform and thus hinder reliable information extraction for either task [15]. Conventional estimation and detection techniques designed for standalone radar or communication systems are generally unsuitable due to the joint waveform and intertwined signal components, necessitating novel signal processing algorithms for accurate target detection and data demodulation. A pragmatic solution involves temporal separation of sensing and communication operations, where well-designed ISAC protocols allocate sensing echoes and uplink data to distinct time slots, thereby reducing interference at the receiver. However, this approach introduces trade-offs in resource utilization and latency, highlighting the need for efficient,

low-complexity, and resource-aware receiver solutions capable of real-time joint processing of sensing and communication signals.

2. **Practical Signal Processing for ISAC**: Although significant advances in SP techniques for ISAC have been reported, most designs rely on idealized assumptions such as perfect hardware, unlimited bandwidth, and full waveform flexibility. In practice, communication-centric design approaches based on standardized waveforms (e.g., 5G-NR) face inherent limitations, as restricted bandwidth constrains radar resolution and range, while the high peak-to-average power ratio (PAPR) of modern signals degrades sensing performance, particularly in terms of dynamic range and detection sensitivity [16]. Furthermore, hardware impairments, including quantized phase shifters, nonlinear power amplifiers, and mismatched antenna arrays, introduce distortions that undermine both sensing accuracy and communication reliability. Hence, advancing ISAC technologies necessitates hardware-aware, resource-constrained SP algorithms capable of operating effectively under standardized waveform constraints, non-ideal hardware, and stringent performance demands to enable practical deployment in next-generation wireless networks.
3. **Full Characterization of the ISAC Performance Tradeoff**: A key challenge in ISAC design is the comprehensive characterization of the performance tradeoff between sensing and communication, which inherently forms a multi-objective optimization problem [17]. Existing studies often analyze only the extreme operating points, where either sensing or communication performance is maximized, leaving the exact shape and location of the Pareto frontier—representing the set of optimal tradeoff solutions—largely unknown [18]. Identifying this frontier is essential for understanding the full range of achievable joint performance and for developing adaptive signaling strategies. Moreover, current research remains limited in realistic scenarios such as multi-user and multitarget environments, where interference, resource contention, and diverse QoS requirements introduce additional complexity. Advancing ISAC thus requires rigorous joint analyses that incorporate both estimation-theoretical bounds for sensing accuracy and the capacity region of multi-user communication systems to enable practical deployment in B5G and 6G networks.
4. **Clutter Suppression and Multi-User Interference in ISAC Systems**: A critical SP challenge in ISAC lies in mitigating both multi-user interference in communication and clutter in radar sensing, as both impair receiver performance and hinder reliable joint operation [19]. Multi-user interference arises from simultaneous transmissions over shared time-frequency resources, while clutter refers to undesired backscattered signals from non-target objects (e.g., buildings, terrain, vegetation) that can obscure legitimate targets in complex environments. Conventional clutter suppression often relies on estimation of the clutter covariance matrix, but this requires prior statistical knowledge that may be unavailable or inaccurate. Dedicated sensing signals or RIS-assisted designs can enhance clutter discrimination, while advanced techniques such as space-time adaptive processing (STAP) exploit spatial-temporal DoFs to separate clutter from target returns and improve detection accuracy. Thus, effective clutter suppression

and multi-user interference mitigation demand adaptive, environment-aware SP strategies to satisfy the stringent performance requirements of next-generation ISAC systems.

To overcome the challenges faced by ISAC systems, the integration of RIS has been introduced. This approach effectively addresses the aforementioned issues and provides several key benefits, as outlined below.

2.3.1 Benefits of Incorporating RIS into ISAC Systems

To tackle the aforementioned challenges, RIS have been identified as a key enabler within ISAC architectures. The integration of RIS into ISAC systems offers several advantages, as outlined below [20]:

1. **Boosted Detection in Obstructed Environments**: Integrating RIS into ISAC systems enables stronger detection of targets in environments where direct LoS is blocked. By redirecting and reinforcing signals, RIS improves signal quality (SNR), making it easier to sense hidden or partially obscured objects. Even when a direct path exists, RIS reflections can amplify signal strength, thereby widening the system's coverage area.
2. **Overcoming Urban Obstruction Challenges**: Urban landscapes often hinder the sensing capabilities of conventional systems due to barriers like buildings that block direct paths to objects such as UAVs, vehicles, or pedestrians. RIS offers an effective solution by rerouting signals around these obstructions, improving detection by strengthening the signal pathway to these otherwise unreachable targets [4].
3. **Interference Suppression Using RIS Beam Control**: In environments with multiple signal sources, interference can degrade ISAC performance. RIS can dynamically adjust the direction and phase of reflected signals, mitigating unwanted interference. This is particularly useful even for highly directional systems, such as those in automotive applications, where residual interference may still pose challenges. The RIS-based approach enhances signal clarity and system reliability.

2.4 ISAC with RIS

A growing body of research is focused on integrating RIS with ISAC, leveraging RIS's ability to dynamically manipulate multipath signal phases to reshape the wireless environment. This makes RIS a promising enabler for ISAC, where S&C functionalities coexist or operate jointly. Beyond communication enhancement, RIS is also being explored for sensing-driven applications, particularly in RF localization [18]. In such scenarios, a mobile device estimates its position using signals from

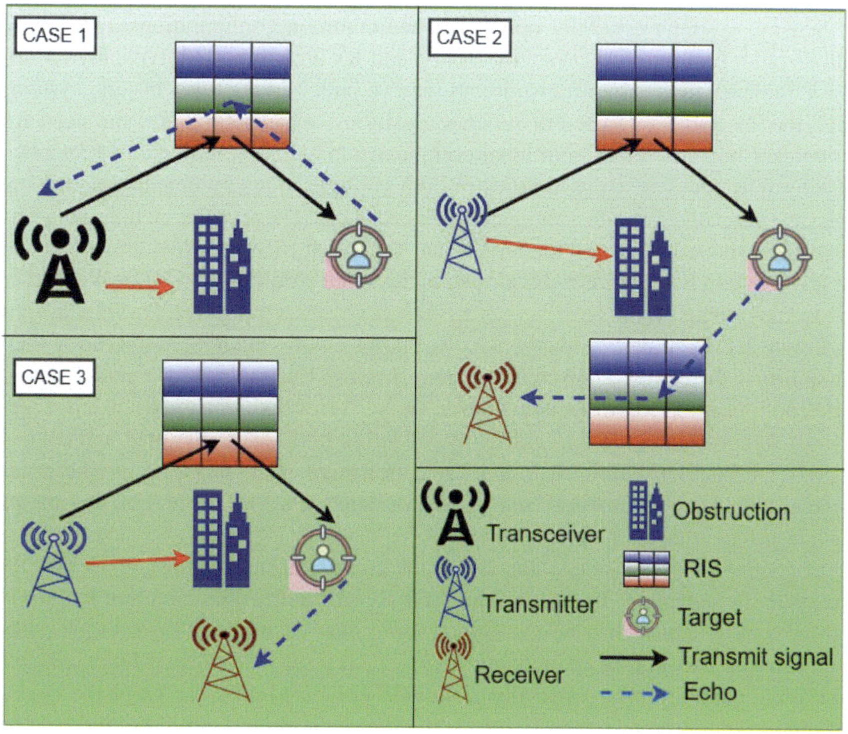

Fig. 2.1 Various sensing scenarios with RIS: Case 1 shows a monostatic setup where a single transceiver handles both transmission and reception. Cases 2 and 3 illustrate bistatic configurations with separate Tx and Rx, highlighting RIS-assisted flexibility in sensing architectures

known terminals, with RIS providing additional reflections to improve accuracy, especially in sparse or NLoS environments. Impressively, RIS can even facilitate localization without traditional infrastructure like access points or BSs. Broader sensing applications include environmental mapping, RF sensing, and radar. The convergence of these capabilities also supports advanced use cases such as spectrum sharing and dual-function systems [21]. The following section focuses specifically on RIS-powered ISAC solutions in various scenarios.

2.4.1 RIS for Communication and Sensing

A myriad work has been reported in the literature where RIS is utilized to enhance the communication by intelligently reflecting the signal incident upon it [22–29]. With the help of tiny meta-materials it beam forms the incident wave to the desired user enhancing the signal quality and coverage [28].

Sensing systems generally operate in two common configurations: monostatic and bistatic. In monostatic systems, the Tx and Rx are collocated, typically requiring a duplexer to separate transmit and receive paths. In contrast, bistatic systems have the Tx and Rx located at separate positions, which eliminates the need for a duplexer but introduces additional complexity in signal coordination. RIS-aided sensing introduces new possibilities in both configurations by enabling additional propagation paths through strategically placed RISs. These paths include both forward RISs (between Tx and the target) and backward RISs (between the target and Rx), which can help redirect signals and echoes that would otherwise be blocked or too weak to detect [18].

Figure 2.1 illustrates various sensing scenarios employing RIS, highlighting their versatility in different system configurations. In Case 1, the setup represents a monostatic configuration, where a single transceiver performs both transmission and reception of signals. Here, the RIS assists by reflecting both the transmitted and echoed signals, the channel exhibits reciprocity, and the design of RIS phase profiles becomes simpler. This enables enhanced sensing performance, even in complex environments such as urban areas with obstructions.

In contrast, Cases 2 and 3 depict bistatic configurations, where the Tx and Rx are spatially separated. In Case 2, the forward RIS helps illuminate the target via indirect paths, which may be the only viable links when direct Tx-to-target paths are obstructed. Despite experiencing double path loss, RIS-assisted links can still contribute meaningfully, especially if the RIS is located near the Tx or the target, potentially equating or exceeding the strength of the direct path. On the Rx side, similar limitations may arise if the direct echo is blocked. In such cases, a backward RIS helps redirect the target's echo to the Rx. However, in the absence of any direct path and full reliance on RISs, the system suffers from quadruple path loss, limiting the overall sensing gain. Case 3 is a simplified version of this bistatic scenario, where only the forward RIS is retained, omitting the backward RIS. While this setup reduces hardware complexity, it also limits the sensing coverage and echo strength. Importantly, in fully bistatic configurations with non-collocated RISs, no duplexer is required, but the phase designs for both RISs must be individually optimized to ensure accurate sensing performance.

These configurations demonstrate how RIS can significantly enhance target detection and localization by intelligently shaping the propagation environment, thereby expanding the applicability of sensing systems in challenging and dynamic scenarios.

2.5 ISAC-RIS Use Cases

In this section, we will highlight the various scenarios or use cases where ISAC-RIS is used. We will touch upon the aspects of ISAC-RIS for UAV systems, ISAC-RIS for IoT systems, ISAC-RIS in V2X scenarios, and ISAC-RIS providing secure communication.

2.5.1 *ISAC-RIS in UAV Systems*

Integrating ISAC into V2X networks enhances vehicular safety, navigation, and real-time decision-making, especially in high-mobility scenarios where GPS alone may fall short. The synergy between S&C supports coordination and low-latency performance, crucial in dynamic environments. Additionally, UAV-enabled ISAC, with its 3D mobility and wide coverage, offers a flexible and power-efficient solution for both S&C in vehicular networks [14]. The two main radio-based UAV sensing paradigms exist: sensing for UAV, which uses sensing to ensure safe flight and monitor low-altitude airspace, and UAV for sensing, where UAVs act as aerial platforms to perform wireless sensing from above.

Mounting lightweight RISs on UAVs enables the creation of virtual LoS links, effectively bypassing ground obstacles and improving coverage in ISAC systems [29]. This combination harnesses the UAV's high mobility and flexible deployment with the RIS's low power consumption, reduced cost, and minimal weight. It is particularly useful for extending coverage to shadowed or hard-to-reach areas. However, UAV movement can introduce Doppler shifts and signal interference, potentially degrading sensing accuracy. To address this, a joint design of RIS reflection coefficients, UAV trajectory, and BS beamforming/processing is essential. Such coordination ensures robust and efficient performance in both communication and sensing tasks.

2.5.1.1 RIS-ISAC-IoT in UAV Systems

In UAV systems, the integration of ISAC-IoT with RIS goes beyond simply merging communication and sensing functions by unlocking new dimensions of aerial intelligence and connectivity [6]. ISAC empowers IoT devices to engage directly in environmental awareness, while RIS enhances this by dynamically shaping EM waves. Beyond signal reflection, intelligent metasurfaces can function as wireless sensors, detecting parameters like temperature, humidity, and object presence. This enables IoT systems to gain deeper situational awareness. Thus, RIS serves not only to boost performance but also as a critical sensing and intelligence enabler within ISAC-IoT networks [30].

Figure 2.2 illustrates an advanced RIS-assisted smart farming system, integrating UAVs, RIS, and ISAC enabling precision farming. The setup includes multiple RISs, both fixed (RIS 1, RIS 2, RIS 3) and integrated (UAV-RIS 1, UAV-RIS 2), strategically deployed to enhance ISAC across a variety of field zones with different agricultural functions.

As per ISAC, radar sensing beams and communication links are utilized to monitor environmental parameters collected from various sensors $S_x \in [1, \ldots, 6]$. For example, a temperature sensor (S_1) and moisture sensor (S_2) are monitored through UAV-assisted links, where UAVs relay radar sensing data to the base station via RIS 1. RISs are easier to deploy on house windows, vehicle glass shield, etc. UAV-RISs also serve to enhance communication coverage in difficult-to-reach zones with the

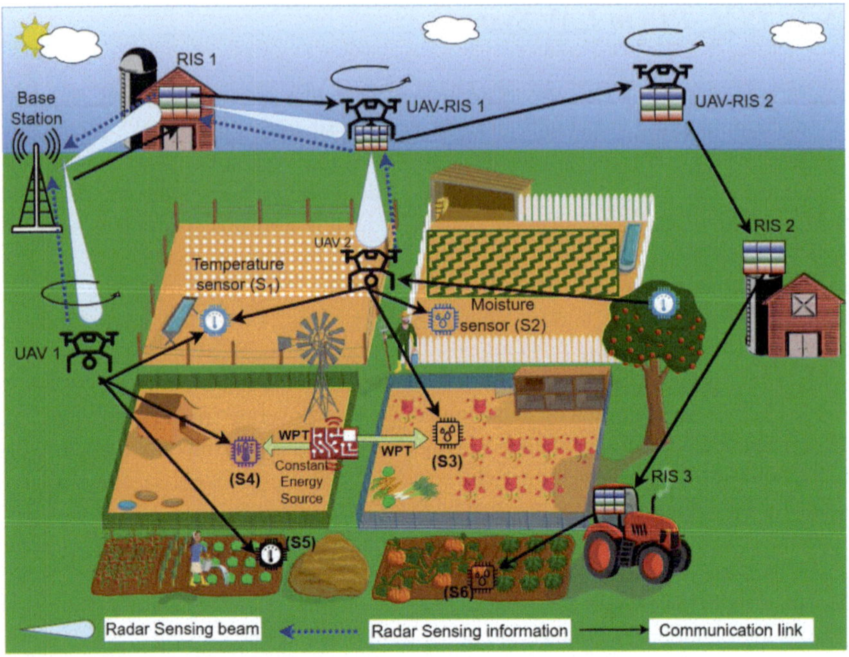

Fig. 2.2 Smart farming scenario with UAV and integrated UAV-RIS performing ISAC to support IoT nodes for communication, sensing, and precision agriculture

help of distributed RIS configuration (RIS 2 and RIS 3), enabling real-time data transmission from sensors like $S_3 - S_6$, which are scattered across crop fields and orchards facilitating efficient control and automation. A constant energy source with wireless power transfer (WPT) supports energy-limited sensor nodes, such as S_3 and S_4, ensuring uninterrupted operation.

This architecture supports both S&C functionalities, enabling timely data collection (e.g., on soil conditions, humidity, temperature) and decision-making for irrigation, harvesting, or fertilization. It can be easily scaled for M RISs, K UAVs, and S sensor nodes. The use of RIS and UAVs significantly improves coverage, reliability, and energy efficiency in rural and dispersed agricultural environments, making it a promising approach for scalable and sustainable smart farming.

2.5.2 *ISAC-RIS in V2X Systems*

ISAC-enabled V2X systems can achieve both high-speed data transmission and precise vehicle localization simultaneously [10]. In V2V scenarios, ISAC enhances environmental awareness, facilitating rapid platooning, secure access, and simultaneous localization and mapping (SLAM). RSUs equipped with sensing capabilities

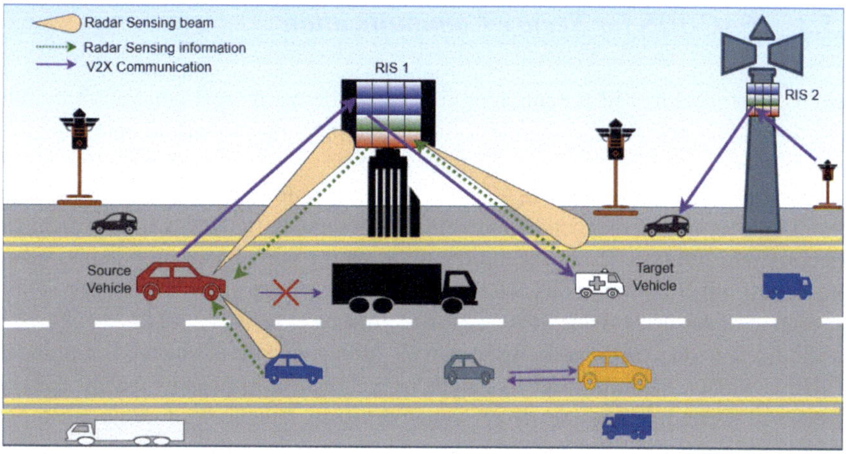

Fig. 2.3 Illustration of RIS-aided ISAC enabling reliable V2X communication for enhanced connectivity and sensing in vehicular networks

can extend a vehicle's perception beyond its direct LoS and field of view. In dense urban areas, where direct links between the base station and users or targets are often obstructed, RIS can be utilized to support ISAC by providing indirect reflective paths.

Figure 2.3 depicts a vehicular ISAC scenario empowered by RIS, showcasing their role in enhancing V2X connectivity and radar-based sensing in complex environments. The setup includes RIS panels (RIS 1 and RIS 2) mounted on an overhead structure like hoardings, roadside watch tower or high-rise buildings nearby road. These RISs help redirect radar sensing beams and facilitate communication between vehicles, particularly in NLoS situations.

In Fig. 2.3, the source vehicle attempts to sense and communicate with a target vehicle. However, a large truck obstructs the direct R&C path. To overcome this, RIS 1 reflects radar sensing beams and V2X communication signals toward the target, enabling the source vehicle to indirectly gather radar sensing information and maintain reliable connectivity. The source vehicle has transceiver antenna, which can transmit sensing beam and gather sensing information as well. RIS 2 similarly assists other vehicles by extending radar coverage and relaying V2X communication signals, particularly for those out of direct LoS from the source.

For those where LoS exists, they continue to communicate without disruption (indicated by purple arrows). By intelligently redirecting ISAC signals, the RISs significantly improve situational awareness, safety, and network reliability for autonomous and connected vehicles navigating real-world road scenarios.

2.5.3 ISAC-RIS for Secure Communication

Secure transmission will remain a significant challenge in next-generation mobile communication networks, as wireless channels inherently provide an open and vulnerable medium where transmitted signals can be easily intercepted by potential eavesdroppers [31], this makes ISAC systems susceptible to malicious threats such as jamming. The presence of active jammers can disrupt legitimate S&C operations by injecting interference. Moreover, unlike traditional communication systems, ISAC integrates both S&C functions, increasing the risk of sensitive information being exposed to adversaries. Thus, ISAC security must be ensured, and RIS can be a better candidate to take on the task. By adaptively tuning the phase shifts and amplitudes of RIS elements, incoming signals can be redirected to form beams toward desired directions, thereby enhancing ISAC signal strength. This beamforming capability improves the anti-jamming SINR, helps mitigate eavesdropping threats, and significantly boosts resistance to jamming. It further strengthens communication security while impacting sensing performance. Integrating an IRS-equipped UAV can significantly improve signal coverage by providing stable LoS links to intended receivers, which plays a crucial role in enhancing communication security. The techniques utilized for ISAC-RIS in UAV environment for enhanced security are listed as follows:

Trajectory-Aware UAV Deployment: The strategic deployment and trajectory optimization of UAVs offer significant advantages for secure data transmission. Leveraging their mobility and flexibility, UAVs can serve as aerial base stations or communication relays, dynamically adjusting their positions to stay closer to legitimate users while distancing themselves from potential eavesdroppers. This adaptability helps to improve the secrecy rate between transmitters and receivers. An RIS-UAV equipment can also support a ground base station by locating receivers and acquiring their CSI. With this information, the joint optimization of UAV flight paths and RIS phase shifts can significantly strengthen the security performance of ISAC system.

Friendly Jamming: By injecting artificial noise (AN) or intentionally generating friendly interference within the null space of legitimate wireless channels, the quality of the eavesdropping link can be significantly degraded, thereby improving the security of the ISAC system. In practical scenario, an RIS can be mounted on the walls and facade of a building, while a UAV transmits AN to interfere with potential eavesdroppers. The coordinated optimization of RIS phase shifts and UAV positioning enables spatially targeted AN transmission and simultaneously enhancing the sensing performance and suppressing eavesdropping threats.

Intelligent Power Control: Wireless signals with insufficient transmit power are highly vulnerable to jamming and noise, resulting in higher bit error rates and degraded communication quality. Conversely, transmitting with excessive power may lead to signal leakage, increasing the risk of interception by unauthorized entities. To maintain secure and efficient transmission, transmit power should be dynamically adjusted based on real-time signal-to-interference-plus-noise ratio (SINR) monitoring. This can be done by designing a UAV flight trajectory, combined with optimized power allocation, such that IRS-based passive beamforming remains in synergy. This coordination enables targeted signal delivery while generating interference toward potential eavesdroppers, thereby enhancing both the sensing accuracy and communication security of ISAC systems.

2.6 Problem Formulation to Improve ISAC System Connectivity Through RIS

We investigate an RIS-assisted bistatic ISAC framework comprising two BSs, an RIS, a single-antenna communication user (CU), and a sensing target capable of transmission using a single antenna. The transmitting BS is equipped with uniform linear arrays (ULAs), consisting of N_t antennas. Under the downlink passive sensing mode considered, one BS functions exclusively as the Tx while the other serves as the dedicated Rx.

As depicted in Fig. 2.4, a direct LoS path between the Tx and target is obstructed. To address this, an RIS comprising K reflecting elements is strategically positioned to establish virtual LoS links for both communication and sensing purposes. The ISAC signal originates at the Tx and is reflected by the RIS. Through the application of active beamforming at the Tx and passive beamforming at the RIS, the communication beam is steered toward the CU, while the sensing beam is directed toward target. The sensing information, resulting from the target's reflection, is subsequently captured and processed at the Rx. It is assumed that perfect channel state information (CSI) is available for both the CU and target. Furthermore, synchronization between the Tx and Rx is maintained, allowing for real-time information exchange.

2.6.1 *Channel Model*

In this subsection, we discussed the channel models for the communication and sensing for considered RIS-aided ISAC system.

2.6.1.1 Communication Model

The signal received at the CU can be modeled as

$$y_c = \mathbf{g}_c \boldsymbol{\Theta} \mathbf{H} \mathbf{x} + n_c, \tag{2.1}$$

where $\mathbf{g}_c \in \mathbb{C}^{1\times K}$ characterizes the RIS-to-CU channel, $\mathbf{H} \in \mathbb{C}^{K\times N_t}$ represents the Tx-to-RIS channel, modeled by Rician fading with Rician factor β, and $\boldsymbol{\Theta} \in \mathbb{C}^{K\times K}$

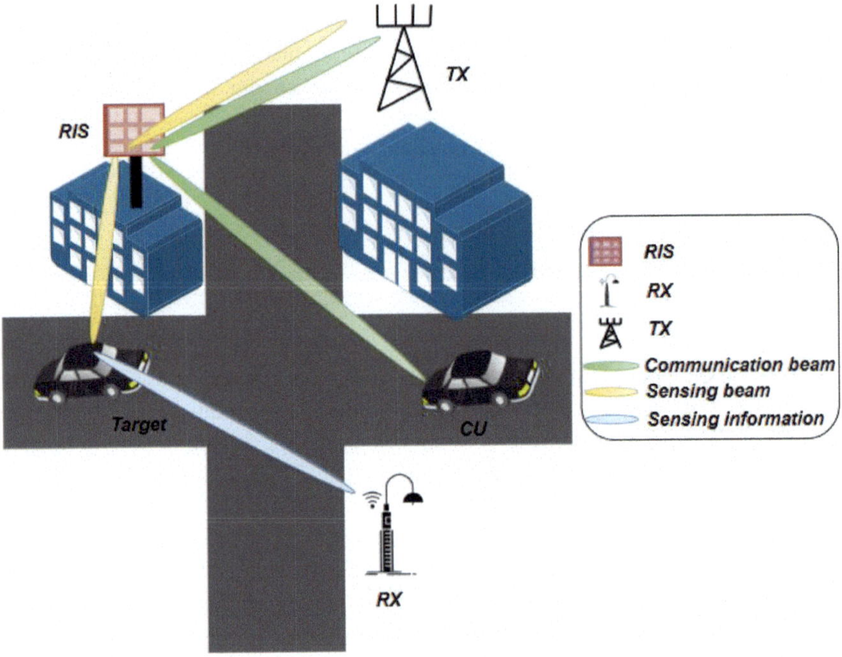

Fig. 2.4 RIS-aided bistatic ISAC system with separate multi-antenna transmitter (Tx) and single-antenna receiver (Rx) enabling simultaneous sensing and communication in vehicular scenario

denotes the diagonal RIS phase shift matrix. The noise term $n_c \sim \mathcal{CN}(0, \sigma_c^2)$ accounts for additive white Gaussian noise (AWGN) at the CU.

Assuming equal gain transmission from all antennas, the combined transmit signal for simultaneous communication and sensing is expressed as

$$\mathbf{x} = \hat{\mathbf{p}} \left(\sqrt{P_B} x_c + \sqrt{P_B} x_s \right), \tag{2.2}$$

where $\hat{\mathbf{p}} = \sqrt{\frac{1}{N_t}} [1, 1, \ldots, 1]^T \in \mathbb{C}^{N_t \times 1}$ is a unit-norm beamforming vector [25, 32], P_B is the total transmit power, $x_c \sim \mathcal{CN}(0, 1)$ represents the information-bearing signal intended for the CU, while x_s is a pseudo-randomly generated sensing waveform that remains uncorrelated with x_c.

The received SINR at the CU is given by

$$\gamma_c = \frac{P_B |\mathbf{g}_c \boldsymbol{\Theta} \mathbf{H} \hat{\mathbf{p}}|^2}{P_B |\mathbf{g}_c \boldsymbol{\Theta} \mathbf{H} \hat{\mathbf{p}}|^2 + \sigma_c^2}. \tag{2.3}$$

2.6.1.2 Sensing Model

The signal received at the radar Rx is modeled as

$$\mathbf{y}_r = \alpha\zeta_t\zeta_r\mathbf{b}(\phi_t)\mathbf{\Theta}\mathbf{H}\mathbf{x} + \mathbf{n}_s, \tag{2.4}$$

where $\mathbf{n}_s \sim \mathcal{CN}(\mathbf{0}, \sigma_s^2)$ denotes the noise, α is the radar cross-section (RCS), ζ_t is the path-loss components for the RIS-to-target link, and ζ_r is the path-loss components for the RIS-to-source link.

The steering vectors $\mathbf{b}(\phi_t) \in \mathbb{C}^{1\times K}$ are given by

$$\mathbf{b}(\phi_t) = \left[1, e^{-j\frac{2\pi d_t}{\lambda}\sin\phi_t}, \ldots, e^{-j\frac{2\pi(K-1)d_t}{\lambda}\sin\phi_t}\right], \tag{2.5}$$

where d_t denote the element spacings and λ is the carrier wavelength.

The sensing SINR is expressed as

$$\gamma_r = \frac{P_B|\alpha\zeta_t\zeta_r\mathbf{b}(\phi_t)\mathbf{\Theta}\mathbf{H}\hat{\mathbf{p}}|^2}{P_B|\alpha\zeta_t\zeta_r\mathbf{b}(\phi_t)\mathbf{\Theta}\mathbf{H}\hat{\mathbf{p}}|^2 + \sigma_s^2}. \tag{2.6}$$

This SINR metric is critical for evaluating the performance of sensing tasks, including target detection and parameter estimation.

2.6.2 Performance Analysis

In this subsection, the performance of the RIS-aided ISAC system is analyzed.

2.6.2.1 Outage Probability (OP)

The proposed RIS-aided ISAC system is deemed to be in outage if the instantaneous SINR of the optimal link drops below the threshold value γ_{th}. The OP of the integrated relay-RIS system is expressed as

$$\bar{P} = \Pr[\gamma_t < \gamma_{\text{th}}] = F_{\gamma_t}(\gamma_{th}), \tag{2.7}$$

where $\gamma_{\text{th}} \in [\gamma_{\text{th,c}}, \gamma_{\text{th,r}}]$ is the threshold SNR. By setting $\gamma_t = \gamma_c$ and $\gamma_t = \gamma_r$, we obtain the OP for the communication and sensing model, respectively.

The OP analysis for integrated RIS-aided ISAC system is shown in Figs. 2.5 and 2.6. The considered simulation parameters are shown in Table 2.1. Specifically, Fig. 2.5 presents the variation in OP for CU with respect to transmit power, for different number of RIS elements. The plots show that the ISAC communication users' OP decreases with an increase in the number of RIS elements. For OP of

Table 2.1 System simulation parameters

Parameter	Value
Carrier frequency f_c	3 GHz
Distance from Tx to RIS d_{tr}	200 m
Distance from RIS to CU d_{ru}	200 m
Distance from Tx to target d_{tt}	200 m
Distance from RIS to target d_{rt}	200 m
Distance from target to Rx d_{tr}	20 m
Rician factor β	1 dB
RCS of the target α	1
Transmit power $P_{\max}$	30 dBm
Noise power at CU σ_c^2	10 dBm
Noise power at Rx σ_s^2	−40 dBm
Number of transmitter antennas N_t	4
Number of receiver antennas N_r	1
Number of IRS elements N	16, 64, 128
Communication SINR threshold $\gamma_{\text{th,c}}$	−10 dB
Rx SINR threshold $\gamma_{\text{th,r}}$	−20 dB

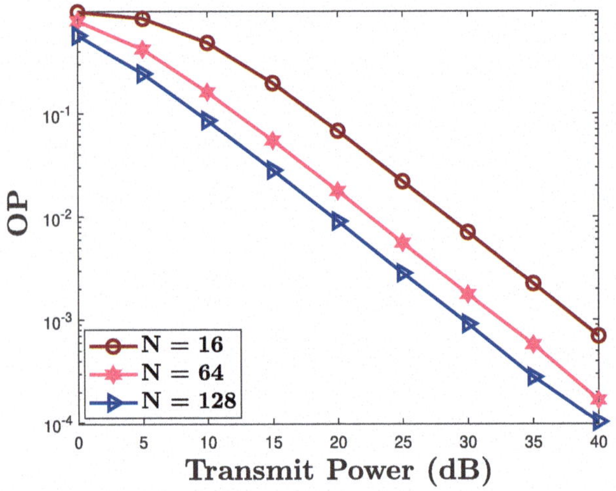

Fig. 2.5 OP versus transmit power for RIS-aided ISAC communication link for various number of RIS elements N values

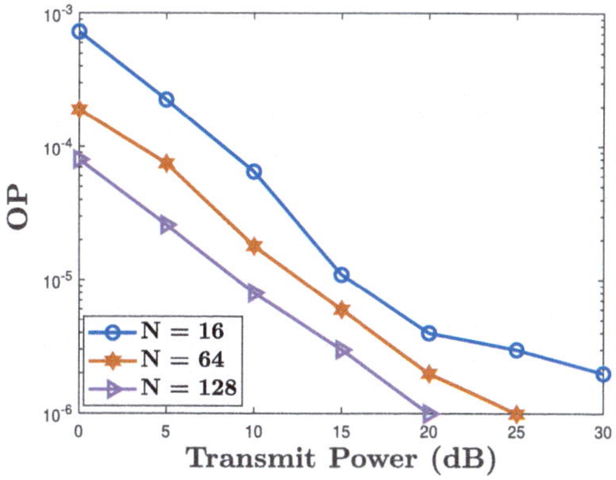

Fig. 2.6 OP versus transmit power for RIS-aided ISAC sensing link for various number of RIS elements N values

10^{-2}, 10 dB transmit power reduction is observed by changing RIS values from $N = 16$ to $N = 128$.

Similarly, in Fig. 2.6, the variation of OP for sensing target with the transmit power for varying numbers of RIS elements is shown. The results demonstrate the efficient integration of RIS in ISAC for target sensing, particularly in terms of OP. For OP of 10^{-5}, the reduction in 4 dB transmit power is observed by changing the RIS values from $N = 16$ to $N = 128$.

2.6.3 *Optimized Outlook*

In this subsection, we will look upon the optimization problem formulation to maximize certain system output with given QoS constraints and provide its optimal solution.

The data rate plays an important role in 6G era. For the given system model in Fig. 2.4 and considering same system equations and channels, we aim to maximize the sum data rate of communication user and sensing Rx for optimized value of transmit power, each having bandwidths of B_c and B_r, respectively. Mathematically given as

$$\text{(P1)}: \underset{P_B}{\text{maximize}} \quad B_c \log_2(1+\gamma_c) + B_r \log_2(1+\gamma_r) \tag{2.8}$$

$$\text{subject to}: \quad (C1): \; \gamma_c \geq \gamma_{\text{th,c}}, \tag{2.9}$$

$$(C2): \; \gamma_r \geq \gamma_{\text{th,r}}, \tag{2.10}$$

$$(C3): \; 0 < P_B < \mathcal{P}. \tag{2.11}$$

The constraints (C1) and (C2) ensure that the desired SINR is obtained at communication user $\gamma_{\text{th,c}}$ (2.3) and Rx $\gamma_{\text{th,r}}$ (2.6), in order to avoid system going into outage. Constraint (C3) ensures transmit power P_B is within the allowable power limit of $\mathcal{P}$.

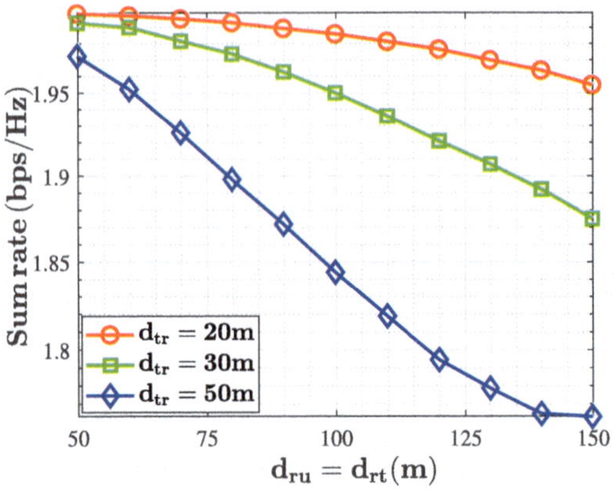

Fig. 2.7 Variation of sum rate with separation distance of RIS-target (d_{rt})/RIS-user (d_{ru}) with LoS distance of Tx-RIS (d_{tr})

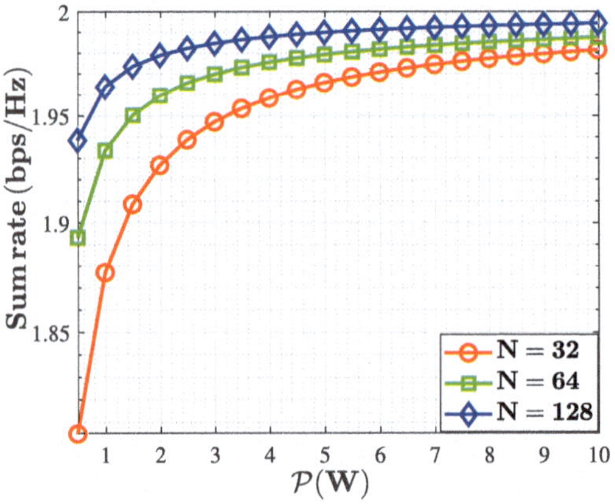

Fig. 2.8 Variation of sum rate with maximum allowed transmit power ($\mathcal{P}$) for different RIS element counts (N). It can be seen that sum rate is higher for more RIS elements and higher $\mathcal{P}$

The suitable solution for this problem can be obtained by using CVX convex solver or MATLAB *fmincon* nonlinear solver utilizing interior point algorithm [27]. We will use the same simulation parameters as given in Table 2.1, except the noise power for CU and Rx ($\sigma_c^2 = \sigma_r^2 = -70$dBm), thresholds $\gamma_{\text{th,c}} = \gamma_{\text{th,r}} = -20$ dB, and maximum allowed transmit power $\mathcal{P} = 5$ W.

It is worth noting that normalized sum rate with bps/Hz unit for equal bandwidths of user and Rx, i.e., $B_c = B_r = B$. To study the impact of system parameters like RIS element number, separation distance, and total number of base station antennas on sum rate, we solve the (P1) and give comparative plots for the same. In Fig. 2.7, the sum rate is varied for separation distance of RIS to target vehicle/RIS to user (both are assumed to be at same distance from RIS). This variation is showed for the various Tx-RIS distances d_{tr}. As both distances d_{rt} and d_{ru} increase, the sum rate decreases due to reduced channel gains.

Figure 2.8 depicts varying sum rate for maximum allowed transmit power $\mathcal{P}$ in Watts with different total RIS elements count $N \in [32]$. It is intuitive that both RIS channels for S&C grow stronger with increasing elements and hence higher sum rate.

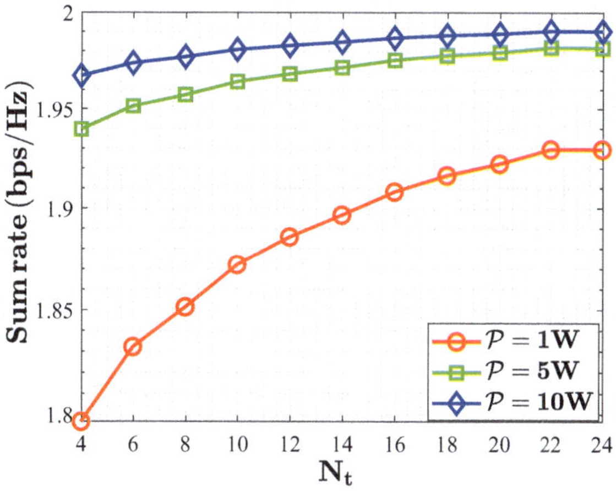

Fig. 2.9 Variation of the sum rate with the number of transmit antennas N_t under maximum transmit power constraint $\mathcal{P}$. The sum rate increases with the number of transmit antennas

Though due to presence of interference term, this change is not much pronounced. A similar observation can be seen in Fig. 2.9, where sum rate is varied for various antenna numbers N_t at BS. Higher antennas means focused beam from BS to RIS, leading to higher rate for both S&C paths and thereby increasing sum rate. From both plots, Figs. 2.8 and 2.9, sum rate increases with $\mathcal{P}$.

2.7 Challenges Faced by ISAC-RIS

In this section, we will discuss the challenges faced by practical ISAC-RIS integrated systems.

2.7.1 Practical Channel Modeling

Accurately modeling the radar echo channel is crucial, as varying assumptions regarding echo propagation paths lead to different radar signal models and processing algorithms [33]. A common assumption is that all returned signals, regardless of their paths, contribute equally to target detection overlooking the impact of propagation delays introduced by reflections. This assumption can be overly idealistic, as such delays may result in ghost targets and introduce ranging ambiguities. Furthermore, the potential for non-negligible interference directly reflected by the RIS or from objects mounted on it toward the radar receiver remains an open question. A promising approach to mitigate such interference is to leverage the quasi-static nature of the BS-RIS channel, whose time-invariant characteristics may allow for statistical suppression of interference. These considerations highlight the necessity for conducting experimental studies and developing data-driven channel models based on real-world ISAC system prototypes.

2.7.2 Deployment and Control of RISs

Many existing studies assume a predetermined RIS deployment configuration, wherein the placement, the number of reflecting elements, and the associated control signaling are fixed and known a priori. However, in practical heterogeneous network environments, the optimal deployment of RISs—including their physical locations, dimensional specifications, and control architectures—plays a critical role in achieving desirable system performance and high operational efficiency [34]. Therefore, dynamic and context-aware RIS deployment strategies must be carefully investigated to adapt to diverse and evolving network conditions.

2.7.3 Implementation and Hardware Constraints

The implementation of ISAC-assisted robotic possesses significant design and programming challenges including (a) on the design front, the primary concern lies in developing ISAC-assisted solutions that accommodate the specific requirements of robotic systems, which differ significantly from other systems in terms of size, shape, weight, and programming capacity. Hence, the dual technology integration must be compact, lightweight, and energy-efficient while maintaining compatibility with existing architectures [35]. Additionally, the formulation of standardization and interoperability is expected to further complicate designing constraint, requiring alignment with evolving global standards to ensure seamless integration, (b) from a programming perspective, the vast amount of data generated from the high-resolution ISAC units necessitates efficient data processing, often reliant on edge or distributed computing architectures. This places additional burden on robotic systems, making it crucial for software developers to balance these demands without overburdening the system. Indeed, sensing accuracy, in general, depends on the operating bandwidth, making it essential to utilize the high frequency, e.g., mmW and terahertz frequencies. However, this introduces cross-band coexistence and related challenges; (a) high-frequency signals require advanced signal processing techniques and high-gain antenna designs due to their poor transmission characteristics. Additionally, developing compact, lightweight, and power-efficient transceivers capable of handling multiple frequencies is particularly challenging, especially for small-scale robotic platforms, (b) besides, distinct frequency bands possess varying hardware and software requirements, adding complexity to system integration and maintenance efforts.

2.7.3.1 RIS Hardware Requirements and System Design

The hardware design of RIS remains an evolving research domain. Presently, most hardware implementations are predominantly optimized for wireless communication applications. However, recent advancements have introduced hybrid passive–active RIS architectures capable of simultaneous reflection and sensing, aiming to enhance reconfigurability and broaden their utility within ISAC frameworks [35]. A thorough understanding of the distinct hardware requirements emerging from ISAC-specific applications is essential for the development of next-generation, efficient RIS technologies. Furthermore, the deployment and operational management of RISs in ISAC-enabled systems represent critical areas of inquiry. Hardware-level considerations include designing RISs capable of handling dual-function ISAC signals—potentially wideband or multiband—while determining the necessary degree of configurability to achieve programmable coupling between S&C channels. From a system-level perspective, effective RIS management entails strategies for deploying multiple RIS units and coordinating their joint operation to fully exploit their potential in enhancing both sensing accuracy and communication quality in mobile and dynamic environments.

2.7.4 Energy Management

Energy efficiency is another critical challenge in the implementation of robotic systems, particularly ISAC-assisted framework, carrying simultaneous execution of high-precision sensing and low-latency communication; (a) continuous data sensing and transmission exerts additional challenge, straining power resources [34]. Further, the presence of high-resolution sensors, e.g., LiDAR scanning, terahertz imaging, etc., requires significant power, (b) while the coexistence of sensing and communication provides multi-fold benefits, it also increases power consumption as both tasks are usually performed simultaneously/rapid succession. More importantly, the real-time processing demand in robotic applications further complicates energy management.

2.7.5 Security and Privacy

A significant challenge in ISAC systems arises from the potential compromise of data secrecy, particularly when communication signals are superimposed over sensing waveforms. While the fusion of communication and sensing offers enhanced functionality, it concurrently introduces critical security and privacy vulnerabilities [36]. The inherent integration of sensor data and communication streams necessitates the implementation of advanced security mechanisms to mitigate cyberthreats, prevent data breaches, and thwart unauthorized access. Protecting user privacy is vital for fostering trust and ensuring the responsible use of sensitive information, particularly in applications involving personal or confidential data [37]. Given the multidisciplinary nature of ISAC networks, collaborative efforts among industry stakeholders, academic researchers, and regulatory authorities are imperative to develop unified security frameworks, establish best practices, and define robust privacy guidelines. Addressing these challenges is essential not only for maintaining the integrity and confidentiality of ISAC transmissions but also for enabling the trustworthy and scalable deployment of intelligent connectivity solutions across various sectors.

2.8 Future of ISAC with RIS Integration

We will look into the prospective areas in which ISAC-RIS can be included for enhanced connectivity, in following section.

2.8.1 ISAC with Simultaneous Transmitting and Reflecting RIS (STAR-RIS)

A recent advancement in this domain is the development of STAR-RIS, which extends the capabilities of conventional RIS by allowing signals to be both transmitted and reflected through the surface. This dual-mode operation enables full-space coverage, making STAR-RIS a promising candidate for next-generation networks such as 6G, where ubiquitous connectivity and integrated sensing are critical. By dynamically adjusting the EM response of each element, STAR-RIS can support simultaneous communication with users located on both sides of the surface—significantly increasing spectral efficiency and spatial flexibility. Moreover, the integration of sensing capabilities into STAR-RIS opens new opportunities for joint communication and sensing systems, leading to more intelligent and responsive network environments [38].

Hybrid deployment of simultaneous transmitting and reflecting reconfigurable intelligent surfaces (STAR-RIS) integrates both S&C functionalities by spatially differentiating their operational regions. This deployment strategy is primarily categorized into half-space hybrid (HSH) and full-space hybrid (FSH) configurations. In HSH deployment, the two sides of the STAR-RIS are functionally partitioned, one dedicated to sensing and the other to communication, making it especially suitable for scenarios where sensing targets and communication nodes are geographically separated. A representative example is a smart factory, where STAR-RIS mounted on the ceiling facilitates internal communication among machines while concurrently monitoring environmental parameters outside the facility. All three STAR-RIS protocols, energy splitting (ES), time switching (TS), and mode switching (MS), can be employed with the ES protocol enabling concurrent operation, albeit at the cost of more complex interference management and signal separation. Efficient allocation of STAR-RIS resources such as time slots or elements must be dynamically managed based on the application's QoS demands; for example, prioritizing communication during peak operational hours and sensing during inspection intervals. In contrast, FSH deployment allows both functionalities to operate across the entire surface on both sides of the STAR-RIS, offering broader coverage. However, simultaneous operation using the ES protocol may lead to mutual interference between echo signals from opposite sides, necessitating the use of TS protocols to alternate between tasks and mitigate cross-interference. This approach simplifies interference mitigation but imposes strict time synchronization requirements to maintain system integrity and operational efficiency.

2.8.2 *ISAC with Beyond Diagonal RIS (BD-RIS)*

With the rapid evolution of RIS technology, a next-generation architecture known as beyond diagonal RIS (BD-RIS), or "RIS 2.0," has recently gained attention for its enhanced functionality and performance. Unlike conventional diagonal RIS architectures where each element operates independently, BD-RIS introduces inter-element connectivity through tunable impedance components, enabling more flexible and efficient beamforming. One of the most prominent configurations within this paradigm is the fully connected (FC) BD-RIS, in which every pair of RIS elements is interconnected to maximize signal manipulation capabilities [39]. Recent research has highlighted the advantages of FC BD-RIS in ISAC systems. For instance, studies have demonstrated that BD-RIS can significantly outperform traditional designs by offering greater beamforming gains and improved spectral efficiency. Specifically, the FC architecture has shown up to a 15% improvement in system sum rate and a 25–30% reduction in transmit power consumption compared to its diagonal counterparts [40]. These benefits make BD-RIS a promising enabler for ISAC applications, where high-precision sensing and efficient communication must be jointly supported.

2.8.3 *Simultaneous Wireless Information and Power Transfer (SWIPT) for RIS-Aided ISAC (SWIPT-RIS)*

The integration of ISAC with SWIPT presents a promising approach for concurrently serving ground users and sensing targets across terrestrial, maritime, and aerial domains. However, such a design introduces increased complexity, particularly in ensuring user fairness between communication and energy services [41]. SWIPT technology enables dual-functional BSs (DFBS) to simultaneously transmit signals to both information users and energy receivers. Specifically, users that decode information from the DFBS are referred to as communication users, whereas devices that harvest energy from the received signals are termed energy receivers. These energy receivers typically consist of low-power IoT devices that require only a minimal amount of power to meet their QoS requirements. Nevertheless, challenges such as severe channel fading and high interference—particularly for users located far from the DFBS—can compromise the performance of both communication and energy transfer, necessitating increased transmission power to fulfill QoS constraints. To ensure sustainable operation, particularly for power-constrained IoT nodes, the integration of wireless power transfer (WPT) into future 6G networks is essential.

By combining ISAC with WPT, 6G networks are envisioned to support a multifunctional air interface capable of concurrently performing sensing, communication, and energy delivery. With the ongoing advancement of enabling technologies, future 6G infrastructures are expected to natively integrate SWIPT and ISAC, thereby paving the way toward unified wireless platforms with comprehensive functionality [42]. Moreover, to support the massive deployment of IoT devices across diverse environments, it is imperative to extend the coverage of current wireless systems. RIS emerge as a key enabler by establishing virtual LoS paths, thereby enhancing network coverage and facilitating connectivity for a larger number of IoT devices in challenging propagation environments.

2.8.4 Machine Learning Techniques for RIS-Aided ISAC

RIS possess the capability to dynamically manipulate wireless propagation environments, thereby enhancing signal quality in scenarios where LoS communication is obstructed or severely attenuated. This work explores a comprehensive architectural framework that synergistically combines ISAC, RIS, and machine learning (ML) to enable a smart, adaptive, and data-driven communication network. Within this framework, environmental sensing data is collected via embedded devices and utilized to optimize network functionalities such as dynamic resource allocation, beamforming, and interference mitigation. These ML-driven insights are further leveraged by a centralized cloud controller that coordinates transmission strategies across multiple BSs to jointly support communication and sensing tasks. This approach significantly enhances network adaptability and operational efficiency [43].

In practical applications such as autonomous driving, environmental sensing plays a vital role in ensuring both safety and intelligence. Mobile communication infrastructures are progressively incorporating wireless sensing capabilities, driven by ISAC technologies. For instance, MIMO systems have been employed to simultaneously maximize data throughput and localization precision. Similarly, integrated radar-communication systems have been shown to reduce positioning errors while improving spectral efficiency. Despite these advancements, several challenges persist—particularly in scenarios lacking a direct LoS link. Ensuring reliable localization and robust communication under resource constraints remains a critical issue. In this context, ML techniques, especially cooperative learning frameworks and reinforcement learning, offer potential solutions by leveraging data from distributed BSs to enhance system-wide localization accuracy and communication reliability. This aligns with the principles of cooperative localization, which aim to optimize performance through multi-agent collaboration and intelligent data fusion. Ultimately, the integration of ML with ISAC and RIS technologies holds promise for realizing highly adaptive and resource-efficient wireless networks.

2.9 Conclusion

The integration of ISAC with RIS presents a transformative paradigm for the future of wireless systems, particularly in the context of 6G and beyond. In this chapter, we examined the foundational principles of R&C, establishing the rationale behind unifying these traditionally separate functions. By exploring the concepts of ISAC, we have highlighted its potential to improve spectral efficiency, reduce hardware redundancy, and enable a variety of novel applications. We discussed several compelling ISAC use cases across domains such as smart cities, autonomous vehicles, and industrial automation. However, the deployment of ISAC systems is not without challenges, particularly due to the conflicting design objectives and performance trade-offs between sensing accuracy and communication reliability. To address these complexities, we emphasized on the pivotal role of RIS as an enabling technology within ISAC frameworks. We provided detailed benefits of incorporating RIS—such as enhanced beamforming, virtual LoS link creation, and improved spatial diversity—and analyzed its dual functionality in both communication enhancement and environmental sensing. Use cases such as ISAC-RIS-enabled V2X systems were presented to illustrate practical implementations that leverage mobility and elevation for coverage extension and environmental awareness. Furthermore, a problem formulation was introduced to improve ISAC connectivity via RIS, supported by channel modeling, performance analysis, and optimization strategies. Despite the promising benefits, ISAC-RIS systems still face several technical challenges, including accurate CSI acquisition, joint optimization complexity, hardware impairments, and dynamic environmental adaptability. Nonetheless, with ongoing research in areas like SWIPT, machine learning, cooperative localization, and intelligent control, the future of ISAC with RIS integration is poised to redefine wireless connectivity. The fusion of ISAC and RIS technologies opens up unprecedented opportunities for creating multifunctional, intelligent, and highly adaptive wireless systems. This integrated approach not only addresses the limitations of conventional architectures but also sets the foundation for next-generation networks capable of simultaneous communication, sensing, and power provisioning.

References

1. Liu, F., et al.: Seventy years of radar and communications: the road from separation to integration. IEEE Signal Process. Mag. **40**(5), 106–121 (2023)
2. Liu, F., et al.: Integrated sensing and communications: toward dual-functional wireless networks for 6G and beyond. IEEE J. Sel. Areas Commun. **40**(6), 1728–1767 (2022)
3. Zhang, A., Rahman, M.L., Huang, X., Guo, Y.J., Chen, S., Heath, R.W.: Perceptive mobile networks: cellular networks with radio vision via joint communication and radar sensing. IEEE Veh. Technol. Mag. **16**(2), 20–30 (2021)
4. Kokare, M.B., Swaminathan, R., Gautam, S.: Performance analysis and optimization with deep learning assessment of multi-IRS-aided IoV network. IEEE Internet of Things J. **12**(9), 11581–11599 (2025)
5. Sturm, C., Wiesbeck, W.: Waveform design and signal processing aspects for fusion of wireless communications and radar sensing. Proc. IEEE **99**(7), 1236–1259 (2011)
6. Cui, Y., Liu, F., Jing, X., Mu, J.: Integrating sensing and communications for ubiquitous IoT: applications, trends, and challenges. IEEE Net. **35**(5), 158–167 (2021)
7. IEEE 802.11bf Task Group: WiFi Sensing Uses Cases (2020). https://mentor.ieee.org/802.11/dcn/20/11-20-1712-02-00bf-wifi-sensing-use-cases.xlsx
8. Huang, Q., Chen, H., Zhang, Q.: Joint design of sensing and communication systems for smart Homes. IEEE Netw. **34**(6), 191–197 (2020)
9. Rao, S.K., Prasad, R.: Impact of 5G technologies on industry 4.0. Wirel. Pers. Commun. **100**(1), 145–159 (2018)
10. Cheng, X., Duan, D., Gao, S., Yang, L.: Integrated sensing and communications (ISAC) for vehicular communication networks (VCN). IEEE Internet of Things J. **9**(23), 23441–23451 (2022)
11. Du, Z., et al.: Toward ISAC-empowered vehicular networks: framework, advances, and opportunities. IEEE Wirel. Commun. **32**(2), 222–229 (2025)
12. Lasi, H., Fettke, P., Kemper, H.G., Feld, T., Hoffmann, M.: Industry 4.0. Bus. Inf. Syst. Eng. **6**(4), 239–242 (2014)
13. Schedl, D.C., Kurmi, I., Bimber, O.: An autonomous drone for search and rescue in forests using airborne optical sectioning. Sci. Robot. **6**, 55 (2021)
14. Pang, X., Guo, S., Tang, J., Zhao, N., Al-Dhahir, N.: Dynamic ISAC beamforming design for UAV-enabled vehicular networks. IEEE Trans. Wirel. **23**(11), 16852–16864 (2024)
15. Hu, J., Valiulahi, I., Masouros, C.: ISAC receiver design: a learning-based two-stage joint data-and-target parameter estimation. IEEE Wirel. Commun. **13**(8), 2105–2109 (2024)
16. Zhang, J.A., et al.: An overview of signal processing techniques for joint communication and radar sensing. IEEE J. Sel. Top. Signal Process. **15**(6), 1295–1315 (2021)
17. Xiong, Y., Liu, F., Cui, Y., Yuan, W., Han, T.X., Caire, G.: On the fundamental tradeoff of integrated sensing and communications under gaussian channels. IEEE Trans. Inf. Theory **69**(9), 5723–5751 (2023)
18. Chepuri, S.P., Shlezinger, N., Liu, F., Alexandropoulos, G.C., Buzzi, S., Eldar, Y.C.: Integrated sensing and communications with reconfigurable intelligent surfaces: from signal modeling to processing. IEEE Signal Process. Mag. **40**(6), 41–62 (2023)
19. Shen, S., Xi, R., Wang, H., Dong, J., Jin, J., Wang, Q.: Clutter suppression based sensing algorithm in ISAC system. In: IEEE 24th International Conference on Communication Technology (ICCT), pp. 1340–1345. China, (2024)
20. Elbir, A.M., Mishra, K.V., Shankar, M.R.B., Chatzinotas, S.: The rise of intelligent reflecting surfaces in integrated sensing and communications paradigms. IEEE Net. **37**(6), 224–231 (2023)
21. Lou, J., Jiang, C., Han, X., Wang, X.: A standardization framework for channel modeling in RIS-assisted integrated sensing and communication systems. IEEE Commun. Stand. Mag. **8**(4), 48–53 (2024)

22. Sharma, N., Kokare, M.B., Swaminathan, R., Gautam, S.: Distributed RIS SWIPT-IoT systems: optimizing spectral efficiency and energy harvesting. IEEE Wirel. Commun. Netw. Conf. (WCNC) 1–6. Italy (2025)
23. Sharma, N., Gautam, S., Kaushik, A., Chatzinotas, S., Ottersten, B.: Fractional programming strategy for rate-energy optimization in RIS-assisted SWIPT IoT networks. In: IEEE Global Communication Conference (GLOBECOM), pp. 5126–5131. South Africa (2024)
24. Sharma, N., Gautam, S., Chatzinotas, S., Ottersten, B.: Integrating multiple RIS for enhanced SWIPT-IoT performance: optimization strategies. IEEE Access **12**, 182841–182855 (2024)
25. Sharma, N., Gautam, S., Chatzinotas, S., Ottersten, B.: Fractional programming based optimization techniques for RIS-assisted SWIPT-IoT system. IEEE Commun. Lett. **28**(12), 2819–2823 (2024)
26. Sharma, N., Gautam, S., Chatzinotas, S., Ottersten, B.: On optimizing RIS-aided SWIPT-IoTs with power splitting-based non-linear energy harvesting. In: IEEE Global Communications Conference (GLOBECOM), pp. 619–624. Malaysia (2023)
27. Sharma, N., Gautam, S.: Optimization strategy for RIS-assisted SWIPT-IoTs with non-linear energy harvesting. In: National Conference Communication (NCC), pp. 1–6. India (2024)
28. Sharma, P.K., Sharma, N., Dhok, S., Singh, A.: RIS-assisted FD short packet communication with non-linear EH. IEEE Commun. Lett. **27**(2), 522–526 (2023)
29. Kokare, M.B., Mishra, R., Sharma, N., Swaminathan, R., Gautam, S.: Performance of integrated relay-RIS in UAV systems: a selection combining approach. In: National Conference Communication (NCC), pp. 1–6 India (2025)
30. Kaushik, A., et al.: Integrated sensing and communications for IoT: synergies with key 6G technology enablers. IEEE Internet of Things Mag. **7**(5), 136–143 (2024)
31. Yang, Z., Zhang, S., Chen, G., Dong, Z., Wu, Y., Benevides da Costa, D.: Secure integrated sensing and communication systems assisted by active RIS. IEEE Trans. Veh. Tech. **73**(12), 19791–19796 (2024)
32. Shambharkar, D., Dhok, S., Singh, A., Sharma, P.K.: Rate-splitting multiple access for RIS-aided cell-edge users with discrete phase-shifts. IEEE Commun. Lett. **26**(11), 2581–2585 (2022)
33. Liu, R., Li, M., Luo, H., Liu, Q., Swindlehurst, A.L.: Integrated sensing and communication with reconfigurable intelligent surfaces: opportunities, applications, and future directions. IEEE Wirel. Commun. **30**(1), 50–57 (2023)
34. Rihan, M., Zappone, A., Buzzi, S., Fodor, G., Debbah, M.: Passive versus active reconfigurable intelligent surfaces for integrated sensing and communication: challenges and opportunities. IEEE Netw. **38**(3), 218–226 (2024)
35. Kaushik, A.: Integrated Sensing and Communications for Future Wireless Networks Principles, Advances and Key Enabling Technologies. Elsevier (2025)
36. Kaushik, A., et al.: Toward integrated sensing and communications for 6G: key enabling technologies, standardization, and challenges. IEEE Commun. Stand. Mag. **8**(2), 52–59 (2024)
37. Wang, F., Li, N., Yang, K., Tao, X.: IRS-aided bi-static ISAC system with security constraint. In: International Conference on Mobile Computing and Network (ACM MobiCom'24), pp. 2222–2229. New York (2024)
38. Saikia, P., Jee, A., Singh, K., Pan, C., Huang, W.-J., Tsiftsis, T.A.: RIS-aided integrated sensing and communication systems: STAR-RIS versus passive RIS? IEEE Open J. Commun. Soc. **5**, 7954–7973 (2024)
39. Esmaeilbeig, T., Mishra, K.V., Soltanalian, M.: Beyond diagonal RIS: key to next-generation integrated sensing and communications? IEEE Signal Process. Lett. **32**, 216–220 (2025)
40. Guang, Z., et al.: Power minimization for ISAC system using beyond diagonal reconfigurable intelligent surface. IEEE Trans. Veh. Technol. **73**(9), 13950–13955 (2024)
41. Kokare, M.B., Gautam, S., Swaminathan, R.: On efficient resource allocation strategies for multi-RIS enhanced V2I SWIPT systems. In: IEEE Wireless Communication Networking Conference (WCNC), pp. 1–6 Dubai (2024)
42. Yaswanth, J., Saikia, P., Singh, K., Kim, Y.H., Duong, T.Q.: Active and STAR-RIS assisted MIMO ISAC systems with SWIPT. IEEE Trans. Cogn. Commun. Netw. **1**, 1–1 (2025)

43. Saikia, P., Singh, K., Huang, W.-J., Duong, T.Q.: Hybrid deep reinforcement learning for enhancing localization and communication efficiency in RIS-aided cooperative ISAC systems. IEEE Internet of Things J. **11**(18), 29494–29510 (2024)

Chapter 3
UDNs: Enhancing Environmental Monitoring with ISAC and NOMA

Vimal Bhatia, Shravan Kumar Namdeo, and Amit Baghel

Abstract This chapter explores the integration of sensing and communication (ISAC) in ultra-dense wireless networks enhanced by non-orthogonal multiple access (NOMA), with a focus on enabling environmental awareness. As next-generation networks evolve to support massive connectivity, low latency, and intelligent decision-making, ISAC emerges as a key enabler by combining radar-like sensing with data transmission capabilities. One of the leading concepts in this competitive environment is the ultra-dense network (UDN), where the access nodes and/or the communication links per unit area are densified, with the aim of improving overall network performance. In ultra-dense network deployments, where interference and mobility pose significant challenges, NOMA offers an efficient solution to support multiple users' tasks over shared resources. This chapter presents the fundamental principles of ISAC and NOMA, their synergies in dense deployments, and the role of joint design in improving network adaptability, energy efficiency, and situational awareness. Use cases such as smart cities, autonomous systems, and the internet of things (IoT) are discussed to illustrate the practical benefits. Challenges and future research directions are also outlined to guide ongoing developments in this emerging area.

Keywords Ultra-dense networks (UDNs) · Integrating sensing and communication (ISAC) · Non-orthogonal multiple access (NOMA) · Environmental monitoring · Internet of things (IoT)

V. Bhatia · S. K. Namdeo (✉)
Department of Electrical Engineering, Indian Institute of Technology Indore, Indore, India
e-mail: phd2401102023@iiti.ac.in

V. Bhatia
e-mail: vbhatia@iiti.ac.in
URL: https://www.iiti.ac.in/

A. Baghel
Department of Electronics & Telecommunication Engineering, Jabalpur Engineering College, Jabalpur, India
e-mail: abaghel@jecjabalpur.ac.in

A. K. Mishra and V. Bhatia (eds.), *Integrated Sensing and Communication for Sustainable Development Goals*, Signals and Communication Technology,
https://doi.org/10.1007/978-981-92-0209-6_3

Table 3.1 Features of different types of small cells [1]

Small cell type	Placement	Coverage	Users	Power	Access
Macrocell	Outdoor	1–50 km	60	60 W	Open
Microcell	Outdoor	200–2000 m	Up to 200	10 W	Open
Femtocell	Indoor	10–30 m	Up to 200	$\leq$100 mW	Open/Closed/ Hybrid
Picocell	Indoor/Outdoor	4–200 m	Up to 64	0.25–2.00 W	Open
Relay	Indoor/Outdoor	Up to 100 m	–	0.25–2.00 W	Open
RRH	Outdoor	Up to 100 m	–	0.25–2.00 W	Open

3.1 Introduction

3.1.1 Background on Ultra-Dense Networks (UDNs)

UDNs involve densely packed small cells and sensor nodes. While they improve spatial reuse and data rates, interference and scheduling are challenges mitigated by non-orthogonal multiple access (NOMA) and coordinated integrated sensing and communication (ISAC). In an UDN, typical high-power macrocells coexist alongside numerous small cells with low power. Combinations of relay nodes, remote radio heads (RRH), femtocells, and picocells can make up these small cells. UDN's adaptability enables the following future network developments: (1) small cells can facilitate site rollout by offering wireless backhaul services as small cells may be installed near end-users; (2) UDN can foster a new ecosystem for the industry. Sixth generation (6G) allows a large variety of services and new use cases; and (3) assistance for both licensed and unlicensed spectra. The widespread installation of small cells has been identified as the optimal solution for boosting the user experience during the past few years as UDN deploys additional cells to use the current spectrum [1]. Figure 3.1 shows ultra-densification as a combination of a microcell, picocell, and femtocell. Generally, UDN small cells can be categorized as fully functional base stations (BSs) (femtocells and picocells) and macro-extended access points (RRHs and relays). A fully functional BS can execute a low-power macrocell's functions within a smaller coverage area. Specifically, a fully functional BS performs the protocol stack's duties [2, 3]. In contrast, a macro-extended access node is an extension for a macro cell that successfully extends its signal coverage and carries out all or only particular physical (PHY) tier duties. In addition, small cells have several characteristics, such as transmit power, coverage, and deployment scenarios [4, 5]. Table 3.1 displays the properties of many small cell types. The many forms of small cells are detailed in the next section.

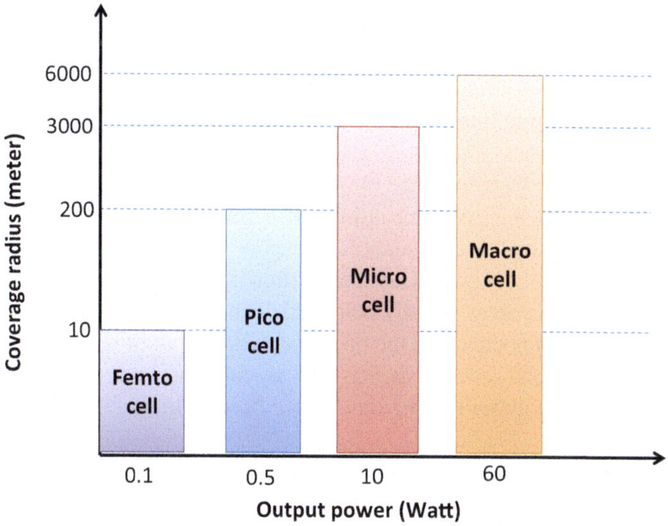

Fig. 3.1 Ultra-densification as a combination of microcell, picocell, and femtocell

3.1.2 Small Cell Types and Fundamental Features of UDNs

Macrocell: Macrocells generally offer a much wider coverage area compared to microcells. Their antennas are typically positioned on tall structures such as towers, rooftops, or other elevated installations, allowing a clear line of sight over surrounding buildings and terrain. The usual transmission power for a macrocell BS is measured in tens of watts. In rural or sparsely populated regions, they are often deployed to maximize coverage across large areas. Owing to their high transmission power, macrocells can serve users over long distances. However, in low-density areas, operators often restrict coverage to control deployment costs. The placement and design of such networks are commonly guided by wireless planning software and validated through on-site measurements.

Microcell: Microcells are a type of small cell designed to provide coverage over distances typically under 2 km. They are often installed in high-density environments such as urban centers, office complexes, and along major highways to boost network capacity and improve service quality. While generally more economical to deploy than picocells, microcells rely on power control mechanisms to limit their coverage range and optimize channel utilization. However, their performance can be more easily affected by sudden changes in traffic demand and by obstacles in the surrounding environment, making them less robust than macrocells in such conditions.

Picocell: Picocells are compact base stations designed to serve limited areas, generally with a coverage radius of up to 100 m. They are strategically positioned in indoor venues or outdoor hotspots to alleviate congestion on macrocell networks. Operating at transmission powers of up to 33 dBm, picocells primarily enhance network capacity rather than extending coverage. Their backhaul connections are typically established via fiber-optic lines or microwave links, similar to those used in macrocells.

Femtocell: Femtocells are compact indoor base stations installed by end-users to improve coverage in small areas, such as homes or offices. They typically support only a limited number of connected devices—often in residential settings—and operate at low transmission power, usually below 20 dBm, with a coverage range of just a few tens of meters. Connectivity to the operator's core network is established via the user's existing broadband service, such as DSL or cable Internet, allowing enhanced indoor signal quality without relying on outdoor base stations.

Relay: Relays are network operator-deployed access points designed to extend coverage to the edges of macrocell service areas and to eliminate dead zones. They typically use wireless backhaul links to connect with macrocell base stations, enabling service in areas where direct coverage is weak or unavailable. While relays and picocells may share similar transmission power levels and coverage ranges, their roles differ: relays primarily function as extensions of macrocells to improve coverage, whereas picocells are mainly deployed to increase network capacity.

RRH: RRHs are radio frequency units installed to extend the coverage of a central base station into distant or hard-to-reach areas. They are linked to the main base station through high-capacity connections such as fiber-optic cables or microwave links. RRHs serve as an alternative to deploying femtocells or picocells when the goal is to enhance coverage density without installing a full base station at the remote site.

3.1.3 Fundamental Features of UDN

To understand UDNs, it is necessary to differentiate them from conventional dense networks. Key distinguishing features include the following:

- **High small-cell density**: UDNs consist of many small cells deployed in close proximity to users, often just tens of meters apart.
- **Idle mode capabilities**: Due to high density, many small cells may remain inactive (idle mode) to reduce interference and conserve power.
- **Interference management**: Proximity of small cells leads to significant inter-cell interference, requiring advanced interference mitigation techniques.

3.1.4 Conventional Cellular Network Architecture

Typical cellular network architecture is a tree network in which BS administrators in the core network oversee each macrocell BS, and all backhaul traffic is delivered to the core network via a specific gateway. A hybrid architecture is recommended for supporting microcell deployments such as femtocell, picocell, and hotspot configurations. The microcell network also follows a tree-based design in this hybrid approach. In this architecture, microcell BS managers maintain each microcell BS in the core network, and the microcell BS backhaul traffic is transferred via broadband Internet or fiber links. Microcellular and macrocellular coverage areas often overlap. Unlike macrocells, microcell BSs can provide high-speed wireless transmission particularly in indoor or hotspot environments. Moreover, both macrocell and microcell BSs are capable of independently managing and transporting user and control data to the connected users. Users can switch between macrocell and microcell coverage zones as needed. This handover (HO) process is coordinated by the macrocells and small cells in the core network. Much research has focused on optimizing HO for 4G and 4G/5G heterogeneous networks (HetNets), introducing various algorithms and strategies to enhance mobility, reduce latency, and improve overall network performance. This architecture is a valuable complement to conventional macrocellular infrastructure, enabling high-speed wireless communication in subregions like indoor or densely populated hotspot areas.

3.1.5 6G Ultra-Dense Cellular Network

In 6G cellular networks, BSs are equipped with large-scale multiple-input multiple-output (MIMO) antenna arrays, incorporating hundreds of antenna elements to support gigabit-level wireless data traffic. To keep the total power consumption in line with that of 5G BSs, the transmit power per antenna element must be significantly reduced. While this approach improves energy efficiency, it also reduces the coverage radius of each BS.

Another fundamental enabler of 6G is the use of millimeter-wave (mmWave) bands, offering ultra-wide bandwidths and the potential to boost transmission capacity by several hundred megahertz. However, mmWave signals are more prone to atmospheric attenuation and blockage, limiting their effective range to roughly 100–200 meters.

The combined limitations of large-scale MIMO and mmWave propagation lead to the need for UDN deployments, with projected densities of approximately 40–50 BSs per square kilometer. This densification creates an ideal platform for ISAC—a key 6G technology that allows the same infrastructure to simultaneously perform high-precision sensing and high-speed communication. Within a UDN, ISAC enables base stations to act as both data hubs and distributed sensors, supporting applications such as environmental monitoring, traffic management, and security surveillance.

As a result, future 6G mobile networks will evolve into ultra-high-density, sensing-enabled infrastructures that require innovative methods for interference coordination, energy management, and cooperative operation among densely deployed BSs.

3.1.6 Energy Efficiency and Ultra-Dense Networks

One of the major challenges in developing next-generation wireless networks is creating solutions that can address the energy limitations introduced by emerging communication paradigms. As the industry moves beyond 5G and toward 6G, energy efficiency (EE) is expected to become a central design priority. Growing awareness of power consumption, environmental impact, and reliance on finite natural resources has driven interest in sustainable, or "green," communication systems.

Sustaining technological progress will require the integration of energy-conscious architectures and operational strategies. This involves investigating approaches such as energy harvesting, bidirectional energy transfer, and intelligent resource management—all aimed at optimizing EE in wireless infrastructures. While these methods hold the potential to reduce both power usage and ecological damage, the notion of "sustainable 6G" remains broad and subject to varied interpretations.

Two promising enablers in this regard are ISAC and UDN. ISAC enables a single hardware platform to perform both communication and sensing, reducing the need for separate infrastructure and lowering overall energy consumption. UDN, when designed with energy-aware resource allocation and sleep-mode mechanisms, minimizes transmission distances and improves spectral reuse, further enhancing EE. Together, ISAC and UDN support applications such as intelligent environmental monitoring, traffic management, and public safety while meeting the dual goals of high performance and sustainability.

- **Energy harvesting and sharing**: Leveraging renewable energy sources and enabling energy transfer or sharing among network nodes to reduce dependence on conventional power supplies.
- **Optimized resource and power management**: Applying techniques such as geometric programming (GP), successive convex approximation (SCA), second-order cone programming (SOCP), and machine learning (ML) models to minimize energy usage without compromising service quality.
- **Energy-conscious sleep and standby modes**: Deactivating or switching idle network elements into low-power or sleep states to conserve energy.
- **ISAC-enabled infrastructure**: Integrating sensing and communication functionalities into the same hardware to reduce the need for separate sensing networks, thereby lowering infrastructure costs and energy consumption.

Although these strategies offer strong technical benefits, their economic practicality is often overlooked. In some cases, hardware modifications or specialized infrastructure requirements can result in higher installation costs and, unexpectedly, greater overall energy consumption. Furthermore, ensuring a consistent supply of renewable energy typically demands substantial upfront investment.

Effective power management remains a critical objective for maintaining user quality of service (QoS), controlling interference, and mitigating pilot contamination. Approaches such as fractional power control, machine learning-driven optimization, and second-order cone programming (SOCP) have shown considerable promise in achieving optimal network utility while balancing performance and energy efficiency.

3.1.7 ISAC: Joint Radar and Communication

Unlike conventional 5G, which primarily focus on delivering communication services, ISAC integrates sensing features that enable communication-assisted innovation services. To highlight the potential of such an ISAC application, three representative scenarios are examined: objective detection and tracking, environmental monitoring, and motion monitoring [6].

- Objective detection and tracking: When user devices are connected to a network, techniques such as positioning and feedback from the user devices can be employed to monitor their location. However, passive objects or targets without network connectivity cannot provide such information directly. In these cases, achieving high precision in estimating both position and velocity becomes essential for tracking devices regardless of their connection status.
- Environment monitoring: In this case, the base station functions as an environmental sensing unit, utilizing existing communication infrastructure. The sensing system can capture data such as rainfall presence, flood levels, crowd density, and traffic conditions. As this information is often time-sensitive, achieving a sensing delay of no more than one minute is crucial.
- Motion Monitoring: This scenario includes applications such as indoor human activity tracking, sleep pattern observation, sports performance monitoring, and gesture recognition. In such cases, reducing false detections is of greater importance than achieving high location accuracy.

ISAC interference management: Moreover, ISAC offers significant potential for optimizing resource usage by merging communication and sensing into a unified system, rather than relying on separate hardware for each. While conventional setups face intra-system interference within communication and sensing individually, combining the two introduces the additional challenges of merging inter-system interference between them. To address this, two main strategies are considered: orthogonal ISAC, which assigns time or frequency resources separately to each functions, and non-orthogonal ISAC, which aims to enhance resource efficiency by allowing both to share time and frequency resources simultaneously.

ISAC waveform design: Waveform design is a crucial aspect of ISAC systems, aiming to create dual-functional signal waveform that supports both sensing and

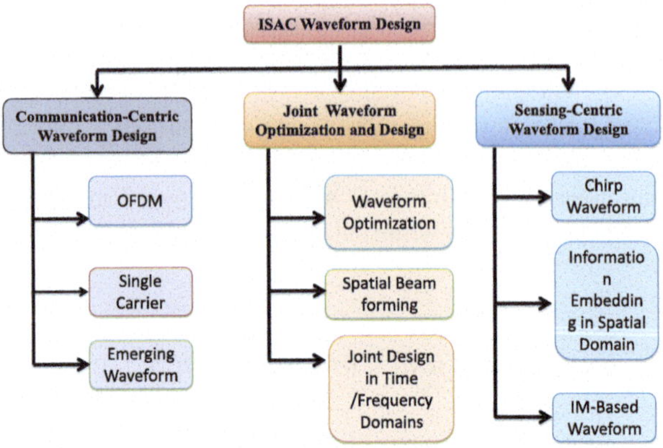

Fig. 3.2 The classification of ISAC waveform design

communication by sharing the same signaling resources. This shared usage enables the realization of integration gains. Based on the degree of integration ISAC waveform ranges from loosely coupled schemes, such as time, frequency, spatial, code-division approaches, to highly integrated design that employs a fully unified waveform (Fig. 3.2).

- Sensing-centric design (SCD)—sensing performance is the primary objective while communication functionality is introduced into existing sensing waveform or infrastructures. A conventional radar waveform by itself does not carry communication signaling; therefore, the SCD approach seeks to embed information-bearing data into the sensing waveform in manner that preserves sensing quality. Let $\mathbf{S}_R$ denote the radar waveform matrix and let $\mathbf{D}$ be the communication data matrix. A sensing-centric ISAC waveform may be formed by an embedding operation $C(\cdot)$ as
$$\mathbf{X} = C\big(\mathbf{S}_R, \mathbf{D}\big). \tag{3.1}$$
- Communication-centric design (CCD)—Communication performance is the principal objective and sensing is introduced as a supporting capability. System design choices—including waveform structure, modulation, and resource allocation—are driven primarily by metrics such as throughput, latency, spectral efficiency, and reliability. Sensing functionality is adapted to fit within the communication-optimized signaling so that quality of service for data transmission is maintained.
- Joint communication and sensing design—In this approach, both sensing and communication functionalities are designed in a fully integrated manner, sharing hardware, spectrum, and signal processing frameworks. Instead of prioritizing one function over the other, the system optimizes performance for both tasks simultaneously. This method achieves higher spectral efficiency and resource utilization, but often involves complex trade-off optimization between sensing accuracy and communication throughput.

3.2 Multiple Access Techniques in ISAC

Incorporating multifunctionality and intelligence into future wireless network design will unlock the full potential of wireless systems, enabling trillions of intelligent devices and users to sense, compute, communicate, harvest energy, and process information seamlessly—anytime, anywhere, and even while in motion [7, 8].

A key challenge, as well as an opportunity, in building such intelligent and multifunctional networks is the need to redefine the concepts of "wireless networks" and "users" in a far broader sense than in the 1G–5G era. In this expanded context, a "user" may not only represent a traditional communication device but also a sensing target, an energy-harvesting terminal, an AI-driven processing node, a training system, or any other entity that the network can serve. Multiple access (MA) techniques form the foundation of wireless network design, playing a central role in serving diverse "users" and efficiently allocating resources such as time, frequency, power, spatial dimensions (e.g., antennas and beams), and signal domains (e.g., messages and codes). In the context of ISAC and UDNs—both key enablers of 6G—MA schemes must handle not only traditional communication demands but also high-precision sensing, energy transfer, and computing tasks. The shift toward intelligent and multifunctional networks introduces new challenges and opportunities for MA design, requiring rethinking resource allocation to ensure optimal coexistence of sensing and communication within dense deployments. To meet these demands, MA strategies must evolve to leverage the synergy between ISAC and UDN, achieving high efficiency, low latency, and robust performance in complex, heterogeneous environments. In earlier generations, TDMA and FDMA dominated 2G, CDMA was central to 3G, and OFDMA combined with SDMA became standard in 4G and 5G. While SDMA-OFDMA has been the primary multiple access (MA) approach for two decades, recent years have seen growing interest in alternatives, particularly NOMA [9, 10]. However, the common binary classification of MA into OMA and NOMA is overly simplistic, grouping diverse schemes under "non-orthogonal" without addressing their differences or design principles [11, 12]. Such a unified design is particularly valuable for 6G and beyond, where diverse services, deployment scenarios, and the integration of technologies like Integrated Sensing and Communication (ISAC) and Ultra-Dense Networks (UDN) demand adaptable, interference-aware MA strategies that can serve sensing, communication, and computation needs efficiently.

3.2.1 OMA

OMA is a basic multiple access method commonly employed in mobile communication networks. It works by partitioning available radio resources—such as frequency bands, time intervals, or codes—into separate, non-overlapping segments that are assigned exclusively to each user. The key idea behind OMA is to avoid interference among multiple users by ensuring they operate in distinct resource blocks. There are four widely recognized OMA methods that have been developed for communication generations from 1G to 4G: frequency division multiple access (FDMA), time division multiple access (TDMA), code-division multiple access (CDMA), and orthogonal frequency division multiple access (OFDMA) [13]. FDMA splits the total spectrum into separate frequency bands, each assigned to a single user. TDMA divides time into discrete slots that are distributed among users. CDMA employs unique orthogonal codes for each user to spread their signals, allowing multiple users to share the same time and frequency resources without interference under ideal conditions. OFDMA further subdivides frequency and time into narrow subcarriers and time slots, grouping them into resource blocks which are then allocated to individual users.

- Advantage
  - **Widespread adoption:** OMA is well-established and extensively utilized in current communication systems, making it easier to facilitate smooth and effortless network expansion.
  - **Simplicity:** OMA simplifies transceiver design, implementation, and management.
  - **Interference free:** OMA prevents interference between users, enabling interference-free transmissions and thereby enhancing the overall quality and reliability of communications. It excels in managing low-to-moderate user loads effectively.

- Disadvantage
  - **Suboptimal spectrum usage**: In OMA, users with low data rates—such as IoT devices needing minimal resources—may still be assigned a full resource block, which can result in inefficient use of the available spectrum.
  - **Limited capacity**: OMA dedicates distinct orthogonal resources to each user, which limits the overall system capacity to the number of available resource blocks. This restriction affects scalability and makes it challenging to meet growing user demands.
  - **Significant signaling overhead**: Improving OMA performance often requires complex user scheduling mechanisms, which increase signaling demands and system complexity.

3.2.2 Power-Domain Non-orthogonal Multiple Access (NOMA)

NOMA employs superposition coding at the transmitter and successive interference cancellation at the receiver to allow multiple users to simultaneously share the same time, frequency, code, or spatial resources. This is achieved by assigning different power levels to each user and superimposing their signals in the power domain, where users with weaker power decode the messages of users with stronger signals. NOMA based on this successive interference cancellation (SIC) technique is known to achieve the capacity region for single-input single-output broadcast channels and capacity for single-input single-output multiple access channels with time-sharing. Extending to multi-antenna systems, MIMO NOMA groups users to exploit spatial multiplexing, managing intra-group interference via SC-SIC and treating inter-group interference as noise. In cases with only one user per group, MIMO NOMA simplifies to SC-SIC, requiring users to cancel all interference. Despite promising theoretical benefits, NOMA has not been incorporated into current wireless standards, largely because its performance improvements over spatial division multiple access and multi-user MIMO have not been sufficiently compelling. Moreover, misconceptions surrounding NOMA remain widespread, prompting further examination of its advantages and disadvantages. ISAC plays a critical role in UDNs by merging sensing capabilities directly with communication infrastructure. Instead of relying on separate sensing systems, ISAC-enabled UDNs utilize shared hardware and spectrum to perform simultaneous sensing and communication tasks. This dual functionality dramatically improves spectral efficiency and reduces hardware cost and energy consumption. In environmental monitoring, ISAC facilitates real-time detection and tracking of environmental parameters such as pollution levels, temperature changes, and hazardous events with high spatial and temporal resolution.

To efficiently manage the massive number of devices in UDNs, NOMA is employed to enhance spectral efficiency beyond traditional orthogonal schemes. NOMA allows multiple users to share the same time-frequency resources by superimposing signals at different power levels and employing successive interference cancellation at receivers. This power-domain multiplexing significantly increases the number of supported devices and reduces latency, making it ideal for dense IoT deployments typical in environmental monitoring networks.

The synergy of ISAC and NOMA in UDNs provides a robust framework to address key challenges:

- **Massive connectivity**: NOMA's power-domain multiplexing supports a large number of low-power sensors distributed across the environment, essential for comprehensive monitoring.
- **Efficient resource utilization**: ISAC's joint sensing-communication operation and NOMA's non-orthogonal access maximize spectrum use, vital in the crowded frequency bands of UDNs.
- **Low latency and high reliability**: Critical for timely environmental alerts and decision-making, ensured by NOMA's efficient resource sharing and ISAC's real-time sensing.
- **Energy efficiency**: Combining sensing and communication hardware in ISAC reduces power consumption, which, along with NOMA's reduced signaling overhead, extends the lifetime of battery-powered sensors.

3.3 System Architecture for Environmental Monitoring

Environmental monitoring systems aim to collect, process, and analyze data from a wide range of sensors distributed across a geographical area to provide timely insights into environmental conditions. The system architecture for such monitoring involves multiple components working together seamlessly to enable real-time, accurate, and scalable data acquisition.

1. **Sensor Layer:** At the foundation are numerous sensing devices deployed throughout the environment. These sensors measure parameters such as temperature, humidity, air quality, pollution levels, radiation, sound, and other ecological variables. In UDNs, sensors are densely deployed to ensure high spatial resolution and comprehensive coverage.
2. **Communication Layer:** This layer handles data transmission from the sensors to centralized processing units or cloud servers. Technologies like ISAC enhance this layer by enabling simultaneous sensing and communication, reducing hardware and spectrum resource consumption. NOMA is often employed here to support massive connectivity, allowing numerous sensors to share communication resources efficiently without orthogonal separation.
3. **Data Processing and Fusion Layer:** Collected data is transmitted to edge or cloud computing resources where it undergoes processing. This includes filtering noise, data fusion from multiple sensors, anomaly detection, and predictive analytics. The goal is to convert raw sensor readings into actionable intelligence.
4. **Control and Management Layer:** This layer manages sensor operation, communication scheduling, power management, and network maintenance. It may implement dynamic resource allocation, user grouping (especially in NOMA scenarios), and interference management to optimize system performance.
5. **Application Layer:** At the top level, processed information is delivered to applications such as environmental dashboards, alert systems, decision support tools, and automated response mechanisms. Users, including environmental agencies, city planners, and emergency responders, access this information for monitoring, planning, and intervention.

Performance Evaluation and Metrics

Performance evaluation is essential to assess the effectiveness, efficiency, and reliability of UDNs utilizing ISAC and NOMA for environmental monitoring. Selecting appropriate metrics helps quantify system capabilities and identify areas for improvement.

1. Sensing Accuracy: This metric measures how precisely the network can detect and quantify environmental parameters such as pollutant concentration, temperature variations, or noise levels. High sensing accuracy ensures reliable data collection for decision-making. Metrics include mean squared error (MSE), detection probability, false alarm rate, and root mean square error (RMSE). In ISAC systems, the joint design of sensing and communication can impact accuracy, requiring performance evaluation that captures this interplay.

2. Communication Efficiency: Communication efficiency reflects how effectively the network transmits sensor data to processing centers. Metrics include spectral efficiency (bits/s/Hz), throughput, and data rate. NOMA's power-domain multiplexing can significantly improve spectral efficiency compared to orthogonal schemes. Evaluating how well NOMA performs in dense deployments with interference and imperfect channel conditions is vital.

3. Latency: Low latency is crucial for timely environmental alerts and responses. Latency measures the delay from data sensing to its delivery and processing. UDNs must minimize latency despite high device density and complex interference environments. Metrics such as end-to-end delay and jitter are commonly used.

4. Energy Consumption and Efficiency: Many environmental sensors are battery-powered, making energy efficiency critical for prolonged operation. This metric evaluates power usage per successful data transmission or sensing task. ISAC's integration reduces hardware redundancy, potentially lowering energy consumption. Additionally, NOMA's ability to serve multiple users simultaneously can decrease transmission time and power requirements.

5. Connectivity and Scalability: Performance is also measured by how many devices the network can support simultaneously without degradation. Connectivity metrics include the number of active users per cell, outage probability, and system capacity. Scalability reflects the network's ability to maintain performance as the number of sensors grows, a key factor in UDNs.

6. Reliability and Robustness: Reliability assesses the system's ability to maintain consistent performance despite channel variations, interference, or hardware failures. Metrics include packet error rate (PER), bit error rate (BER), and system availability. Robustness refers to the network's tolerance to environmental changes and operational disturbances.

7. Security and Privacy Metrics: Given the sensitivity of environmental data, performance evaluation must also consider data security and user privacy. Metrics can include encryption overhead, intrusion detection accuracy, and resilience to attacks.

3.3.1 Challenges and Future Directions

Despite the promising capabilities of ultra-dense networks (UDNs) empowered by ISAC and NOMA for environmental monitoring, several challenges remain. Managing interference in dense deployments, ensuring synchronization among numerous sensors, and maintaining energy efficiency in battery-powered devices require advanced solutions. Additionally, designing scalable algorithms for dynamic resource allocation and user grouping is critical for system performance. Future research may focus on integrating machine learning techniques for adaptive network management, enhancing security and privacy of environmental data, and developing standardized protocols for seamless ISAC-NOMA integration. Addressing these challenges will be key to realizing robust, efficient, and intelligent environmental monitoring systems. Comprehensive evaluation using these metrics, often through simulations and field tests, guides the design and optimization of UDNs with ISAC and NOMA for environmental monitoring, ensuring that the system meets application requirements and operates efficiently in real-world conditions.

3.4 Conclusion

This chapter demonstrates the promising ISAC with NOMA in UDNs. As future wireless systems demand massive connectivity, low latency, and intelligent decision-making, ISAC offers a unified framework by combining radar-like sensing with data transmission. The densification in UDNs enhances network capacity and coverage but raises challenges like interference and mobility management. NOMA effectively addresses these by enabling multiple users to share spectrum simultaneously, improving spectral efficiency.

The synergy between ISAC and NOMA enables networks to adapt dynamically, optimize energy use, and enhance situational awareness, which are essential for applications like smart cities, autonomous systems, and industrial IoT. These use cases highlight the practical benefits of jointly designing sensing and communication capabilities.

Despite the advantages, challenges such as interference management, reliable coexistence of sensing and communication, and hardware constraints remain. Future research is needed to develop advanced algorithms and resource allocation strategies to unlock the full potential of ISAC-enabled ultra-dense networks. Overall, this integration marks a significant step toward next-generation wireless networks capable of supporting diverse, intelligent applications.

3.5 Applications of ISAC-Enabled Ultra-Dense Networks with NOMA

- **Smart Cities**
  Real-time environmental monitoring (air quality, noise, traffic), intelligent lighting and energy management, and enhanced public safety through situational awareness.
- **Autonomous Systems**
  Vehicle-to-everything (V2X) communication and sensing for collision avoidance, drone navigation in urban environments, and robotics automation with integrated sensing and communication.

- **Industrial Internet of Things (IIoT)**
  Predictive maintenance via integrated sensing of machinery health, factory automation with low-latency control, and asset tracking in complex facilities.
- **Healthcare and Remote Monitoring**
  Continuous patient monitoring combining vital sensing and data transmission, and smart hospitals with integrated communication and environmental sensing.
- **Augmented and Virtual Reality (AR/VR)**
  Enhanced user experience with environmental sensing for context-aware content, and low-latency communication supporting immersive applications.
- **Emergency and Disaster Management**
  Rapid deployment of communication and sensing infrastructure during crises, and real-time situational awareness to aid rescue operations.

References

1. Nakamura, T.: 5G Evolution and 6G. In: Proceedings of the IEEE Symposium on VLSI Technology, pp. 1–5 (2020). https://doi.org/10.1109/VLSITechnology18217.2020.9265094
2. Chen, S., Qin, F., Hu, B., Li, X., Chen, Z.: User-centric ultra-dense networks for 5G: challenges, methodologies, and directions. IEEE Wireless Commun. **23**(2), 78–85 (2016). https://doi.org/10.1109/MWC.2016.7462488
3. Wei, Y., Hwang, S.H.: Optimization of cell size in ultra-dense networks with multiattribute user types and different frequency bands. Wireless Commun. Mob. Comput. **2018**, 1–10 (2018)
4. Zhang, H., Song, L., Zhang, Y.J.: Load balancing for 5G ultra-dense networks using device-to-device communications. IEEE Trans. Wireless Commun. **17**(6), 4039–4050 (2018). https://doi.org/10.1109/TWC.2018.2819648
5. Zhong, Y., Ge, X., Yang, H.H., Han, T., Li, Q.: Traffic matching in 5G ultra-dense networks. IEEE Commun. Mag. **56**(8), 100–105 (2018). https://doi.org/10.1109/MCOM.2018.1700956
6. Cui, Y., et al.: Integrating sensing and communications for ubiquitous IoT: applications, trends, and challenges. IEEE Netw. **35**(5), 158–167 (2021)
7. Clerckx, B., Huang, K., Varshney, L.R., Ulukus, S., Alouini, M.S.: Wireless power transfer for future networks: signal processing, machine learning, computing, and sensing. IEEE J. Sel. Topics Signal Process. **15**(5), 1060–1094 (2021)
8. Zhou, W., Zhang, R., Chen, G., Wu, W.: Integrated sensing and communication waveform design: a survey. IEEE Open J. Commun. Soc. **3**, 1930–1949 (2022)
9. Study on downlink multiuser superposition transmission (MUST) for LTE. Technical Report TR 36.859, 3rd Generation Partnership Project (3GPP) (Dec 2015), 3GPP (2015)
10. Study on non-orthogonal multiple access (NOMA) for NR. Technical Report TR 38.812, 3rd Generation Partnership Project (3GPP) (Dec 2018), 3GPP (2018)
11. Clerckx, B., Mao, Y., Jorswieck, E.A., Yuan, J., Love, D.J., Erkip, E., Niyato, D.: A primer on rate-splitting multiple access: tutorial, myths, and frequently asked questions. IEEE J. Sel. Areas Commun. **41**(5), 1265–1308 (2023)
12. Clerckx, B., Mao, Y., Schober, R., Jorswieck, E.A., Love, D.J., Yuan, J., Hanzo, L., Li, G.Y., Larsson, E.G., Caire, G.: Is NOMA efficient in multi-antenna networks? A critical look at next generation multiple access techniques. IEEE Open J. Commun. Soc. **2**, 1310–1343 (2021)
13. Tse, D., Viswanath, P.: Fundamentals of Wireless Communication. Cambridge University Press, Cambridge, U.K. (2005)

Chapter 4
Symbiotic Sensing and Backscatter Communication for the Sustainable Internet of Things

Muhammad Bilal Janjua, Çağrı Özgenç Etemoğlu, Muhammad Sohaib J. Solaija, and Adnan Kavak

Abstract This chapter presents a comprehensive exploration of symbiotic radio, emphasizing its integration with sensing and backscatter communication within the context of ambient power-enabled Internet-of-Things (Ambient IoT) applications. It commences with an examination of the foundational principles of symbiotic radio and extends to the fundamental concepts of sensing and backscatter communication, highlighting their critical roles in promoting energy efficiency and sustainability. The discussion incorporates relevant 3GPP use cases, including sensor data collection, inventory management, command execution, and positioning. Various communication topologies, specifically monostatic and bistatic configurations, are analyzed, complemented by a practical example of sensing-assisted backscatter communication that localizes mobile targets to enhance data transmission efficacy. Furthermore, the chapter addresses signal and system design considerations for employing a unified signal for both sensing and backscatter communication. The concluding section outlines potential applications, use cases, challenges, and future research directions, thereby providing a holistic perspective on the advancements and implications of these technologies in the evolving landscape of Ambient IoT.

M. B. Janjua
R&D Department, Oredata, Esenler, Istanbul, Türkiye

M. B. Janjua (✉) · Ç. Ö. Etemoğlu
R&D Department, Türk Telekom, Üsküdar, Istanbul, Türkiye
e-mail: muhammadbilal.janjua2@partner.turktelekom.com.tr; bilal.janjua@ieee.org

Ç. Ö. Etemoğlu
e-mail: cagriozgenc.etemoglu@turktelekom.com.tr

M. S. J. Solaija
Institute of Defence Technologies, Gebze Technical University, Gebze, Kocaeli, Türkiye
e-mail: solaija@gtu.edu.tr

A. Kavak
Department of Computer Engineering & Wireless Information and Intelligent Systems (WINS) Research Center, Kocaeli University, 41380 İzmit, Kocaeli, Türkiye
e-mail: akavak@kocaeli.edu.tr

A. K. Mishra and V. Bhatia (eds.), *Integrated Sensing and Communication for Sustainable Development Goals*, Signals and Communication Technology,
https://doi.org/10.1007/978-981-92-0209-6_4

Keywords Ambient IoT · Backscatter communication · ISAC · Sustainability · Symbiotic radio · Symbiotic sensing

4.1 Introduction

The use of wireless networks has pervaded all aspects of our society. The applications of these systems range from entertainment to education, commerce to logistics, healthcare to finance, agriculture to industry, utility to transportation, and so on. In essence, continuous advances in wireless communication paradigms, characterized by increasing data rates, reduced latency, and enhanced reliability, allow society to eliminate geographical barriers and empower global collaboration.

One aspect of ubiquitous wireless connectivity is the connectivity of "everything." This builds on the premise of the Internet-of-Things (IoT) paradigm introduced around fourth generation (4G) of cellular networks, which was then further enhanced by the massive machine-type communication (mMTC) service promised by fifth generation (5G). The main idea in both cases was to allow (a very large number of) devices to directly interact with each other. Rather than considering all devices to be powerful and sophisticated, such as smartphones or computers, the IoT paradigm would allow low-end devices like sensors (e.g., temperature) to communicate their information that can be used to automate system parameters (turning the air conditioner on/off) remotely. Some of the realistic applications of IoT include smart homes (lighting, thermostats, etc.), wearable health monitoring systems (glucose monitoring, remote vital signs monitoring), agriculture (livestock monitoring, automated irrigation) as illustrated in Fig. 4.1.

According to an estimate, there were $\approx$ 11 billion devices in 2019, with this number expected to grow up to 30 billion by 2030 [1]. Another source puts latter number closer to 40 billion [2]. Considering the enormous potential of IoT, it is imperative to ensure its sustainability and accessibility throughout the globe. The next-generation systems envisage at least a 10-fold increase in data rates, device density, reliability, and latency. On the other hand, the sustainable development goals (SDGs) laid down by United Nations (UN) in 2015 include reduced inequalities (SDG 10), sustainable cities and communities (SDG 11), and responsible consumption and production (SDG 12) which effectively mean that the energy consumption of wireless networks cannot increase any further than the current situation [3]. This situation necessitates the development of technologies that drastically cut down the energy consumption in wireless networks. One of the candidates being considered is the ambient IoT (A-IoT) under 3rd Generation Partnership Project (3GPP). An A-IoT network harvests energy from ambient sources to power its communication which allows for massive deployment of IoT devices at low operational cost. These devices are often battery-free and harvest energy from natural (e.g., solar) or ambient (e.g., radio waves) sources and are also referred to as "zero-energy" in some cases. Consequently, A-IoT targets the low-end IoT use cases comprising ultra-low complexity devices with extremely low

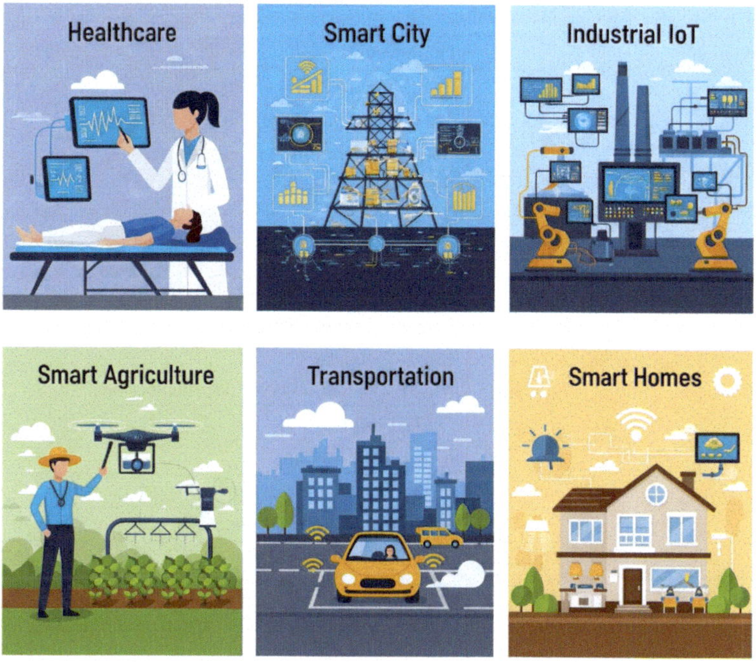

Fig. 4.1 Illustration of IoT applications including, healthcare, smart city, industrial IoT, smart agriculture, transportation, and smart homes

power consumption and small form factors. Till now, 3GPP discussions have revolved around device types, use cases, key requirements and related design challenges [4].

While the fundamentals of energy harvesting and backscatter communication (where data to be transmitted is embedded on top of existing radio signals) are well-established, one of the main concerns regarding A-IoT is how this technology can complement or even coexist with the existing wireless (including but not limited to 3GPP, i.e., cellular) systems. Given that A-IoT relies heavily on the ambient signals/systems, we take a look at how these (A-IoT and environmental) systems can enhance each other's performance in the form of a symbiotic relationship, i.e., how the existing wireless system may be used to support the A-IoT devices. The symbiosis may either be *facultative* (where the entities can survive independently but chose to share resources) or *obligatory* (where one organism in dependent on the other for its survival).

This chapter assumes readers' minimal background knowledge on the topic of A-IoT and symbiotic radio. Consequently, we first provide an overview of the basic concepts related to backscatter communication, wireless sensing, and the potential of a symbiotic relationship between the two in next-generation wireless networks. The list of abbreviations is provided in Table 4.1.

Table 4.1 List of abbreviations

Abbreviation	Definition
2D	Two-Dimensional
3GPP	3rd Generation Partnership Project
4G	Fourth Generation
5G	Fifth Generation
A-IoT	Ambient IoT
FFT	Fast Fourier Transform
FMCW	Frequency-Modulated Continuous Wave
FSK	Frequency Shift Keying
IoT	Internet-of-Things
mMTC	Massive Machine-Type Communication
OOK	On-Off Keying
PSK	Phase-Shift Keying
RF	Radio Frequency
RFID	Radio Frequency Identification
SDG	Sustainable Development Goal
SR	Symbiotic Radio
SSBC	Symbiotic Sensing and Backscatter Communication
UN	United Nations

4.2 Fundamental Concepts

This section presents the fundamental concepts on backscatter communication, wireless sensing and symbiotic radio

4.2.1 Backscatter Communication

In conventional IoT devices, wireless data transfer is facilitated through the generation of a radio frequency (RF) carrier signal at the device. This signal is modulated based on the specific information being transmitted, followed by amplification to ensure that it can effectively travel through the air. However, the processes of carrier signal generation and amplification are notably energy-intensive. This high energy consumption leads to rapid depletion of the IoT device's battery, consequently reducing its operational lifespan. As a result, these devices require frequent management for battery replacement, incurring additional labor and replacement costs.

A passive wireless communication by means of reflected power was proposed by Harry Stockman in 1948 [5]. This method, now referred to as backscatter communication, where the device does not generate its own RF signal; instead, it modulates an external signal by altering its antenna impedance. This modulation can involve changes in amplitude, phase, or frequency, allowing the device to convey information without the need for active signal generation components. The advantages of backscatter communication are significant. By relying on an external carrier signal, this method significantly reduces energy consumption, extending the operational lifespan of IoT devices and minimizing the necessity for regular battery replacements. Consequently, this innovation alleviates the logistical challenges and costs associated with battery management.

4.2.1.1 Backscatter Device Design

A passive backscatter device mainly consists energy harvesting, signal detection, data modulation, and memory parts. A simple architecture of a passive backscatter device is illustrated in Fig. 4.2. Energy is required to power-up and perform the basic functionalities of the backscatter device, which is harvested from the incident carrier signal transmitted by the base station. Particularly, the incident carrier signal is divided into two part, one part is used for energy harvesting and other part is given to the detector for signal retrieving the information from the received signal or modulating it to transmit the information.

Energy Harvesting

In the energy harvesting step a rectifier is used to convert the received RF signal to direct current. This process helps the backscatter devices to power-up the internal circuitry without a separate power source. A power management unit is also installed

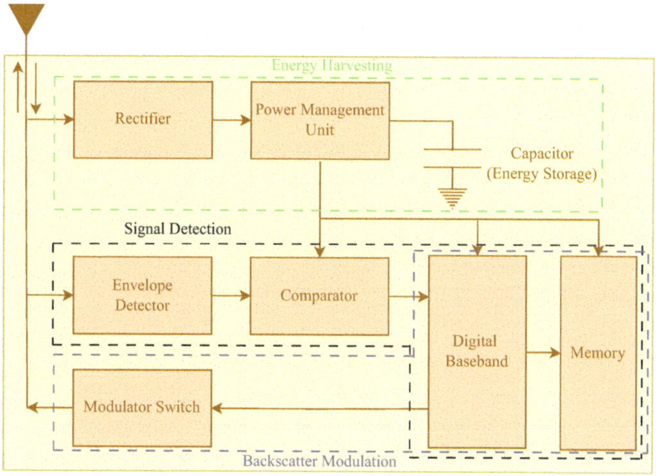

Fig. 4.2 Passive backscatter device architecture. There are three main components of the backscatter device including Energy Harvesting, Signal Detection, and Backscatter Modulation

in the device that manages the energy storage and ensures the voltage levels are suitable for the operation of internal components including comparator, digital baseband and memory.

Signal Detection
To retrieve the information at the backscatter device, the received signal is first passed through an envelope detector, which detects the variation in the amplitudes of the received RF signal and extracts the modulated data that needs to be processed for extracting the encoded information. The output of the envelope detector is given to a comparator that compares it with a predefined threshold. This method helps in determining either the received signal contains data or fall below the threshold. If there is a data in the signal then the signal is sent to the digital baseband for further processing such as digitizing, filtering, and error correction. Afterward, the digitized data is stored in the memory.

Backscatter Modulation
To modulate the information on the incident carrier signal from an external source, the backscatter devices regulates the working state of the antenna through an impedance switch. By varying the load impedance connected to the antenna, the incident carrier signal is modulated. Through controlled variations of the load impedance, the backscatter device performs different modulations such as on-off keying (OOK), phase shift keying (PSK), and frequency shift keying (FSK) and causes the change in the amplitude, phase, and frequency of the carrier signal, respectively, based on the information to be transmitted. For example to transmit binary information with OOK, the backscatter device varies the amplitude of the carrier signal by switching the antenna into non-reflecting and reflecting modes through impedance matching and mismatching, respectively. The reflecting mode represents the information bit "1" and the non-reflecting mode represents the information bit "0" at the receiver.

4.2.1.2 System Design

Figure 4.3 illustrates a simple monostatic single device backscatter communication system. In this configuration, a reader transmits a continuous RF carrier signal x. The backscatter device receives $x(t)$ over the forward channel $h_f(t)$ (i.e., reader to backscatter device). Upon receiving the $x(t)$, the backscatter device modulates it by varying its impedance based on the available information $b(t)$. The reader receives the backscattered signal from the device over the backward channel $h_b(t)$ (i.e., backscatter device-to-base station), which is expressed as

$$y(t) = h_b(t) * (h_f(t) * \alpha b(t)x(t)) + n(t), \tag{4.1}$$

where α is the reflection coefficient of the backscatter device, $n(t)$ denotes the additive white Gaussian noise. The reader cancels out $x(t)$ from the received signal $y(t)$, and performs the demodulation and decoding operations to extract $b(t)$. As the major power consuming and complex operations for communication are performed by the

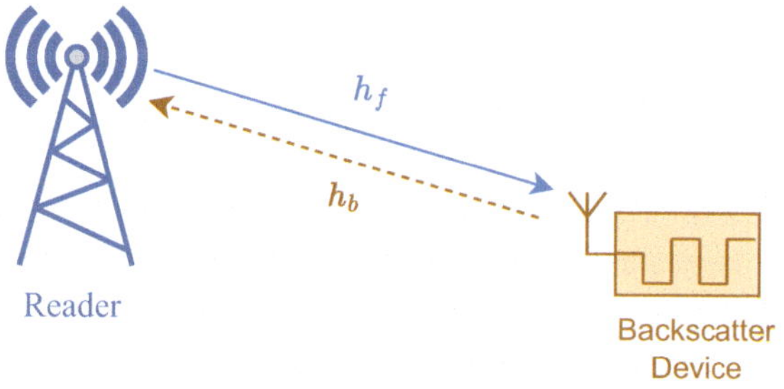

Fig. 4.3 Illustration of monostatic backscatter communication system where both the transmitter and receiver are co-located. The forward (h_f) and backward (h_b) channels are reciprocal in this configuration

reader, backscatter communication presents a promising solution for enhancing the sustainability and efficiency of IoT systems.

The straightforward design of the monostatic backscatter system presents several limitations for the backscatter communication, including restricted range, low data rates, low reliability, and susceptibility to interference (e.g., self-interference). In contrast to conventional active communication, backscatter communication encounters double path loss in both the forward and backward channels. This phenomenon significantly constrains the reliability and effective range of backscatter communication, ultimately impacting its overall performance in practical applications.

These challenges can be addressed by implementing a bistatic backscatter system shown in Fig. 4.4, which separates the transmitter and receiver into separate physical locations. This architectural separation is the key to enhancing communication performance. Specifically, either the transmitter or the receiver can be strategically placed near the backscatter device to significantly reduce path loss. Furthermore, this configuration improves scalability by simplifying the management of multiple tags with reduced contention, enhances reliability, and can mitigate interference from external RF sources.

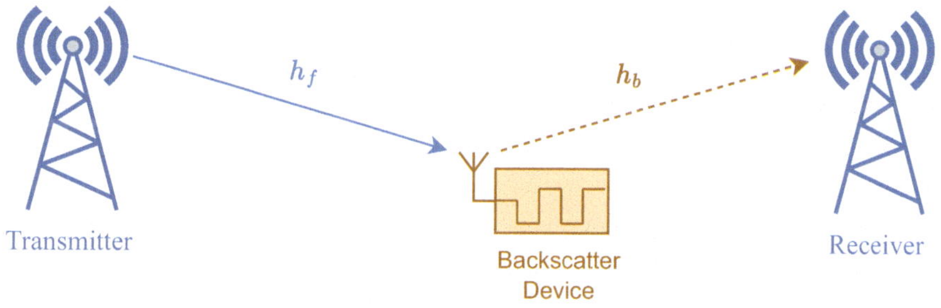

Fig. 4.4 Illustration of bistatic backscatter communication system where the transmitter and receiver are spatially separated. The forward (h_f) and backward (h_b) channels are independent in this configuration

Although bistatic design provide significant advantages, it introduces its own set of significant challenges stemming directly from its separated architecture. The most critical is direct-link interference, where the receiver is overwhelmed by the transmitter's powerful signal path. Additionally, the system must contend with multipath fading and increased design complexity. A fundamental operational requirement is the need for strict synchronization between the transmitter and receiver to ensure accurate signal decoding, especially in deployments with multiple devices.

4.2.2 Wireless Sensing

In simplest possible terms, wireless sensing refers to detection or measurement of physical parameters/conditions using electromagnetic waves. This can be used for a wide variety of applications including military (radar), autonomous driving (collision avoidance), human activity detection (occupancy detection or gesture recognition) and many more. In the following subsections, we will first look at the basic principle of wireless sensing, followed by how the most common physical parameters are typically extracted from the wireless signals.

4.2.3 Principle of Wireless Sensing

Wireless sensing involves sending out known radio signals, referred to as *sensing* signals, in the environment. The interaction of these signals with the objects (targets) in the environment leads to changes in the signal (in terms of power, phase, or frequency). Once this modified signal reaches the sensing receiver, the changes in the signal serve as a source of information regarding the target. This phenomenon is illustrated in Fig. 4.5. The scenario on the left represents the overly simplistic case where the only object present in the environment is the target of interest. However, the scenario on the right indicates the presence of unwanted objects, referred to as *clutter*. In such a case, the sensing receiver (which may be co-located with the transmitter or separately) would receive reflections from both the target and the clutter. Here, it is imperative for the sensing receiver to differentiate between the two and focus on its intended target.

4.2.4 Approaches for Wireless Sensing

Wireless sensing finds its roots in radar (**Ra**dio **D**etection **a**nd **R**anging) which, though developed at the end of the nineteenth century, became popularized during the Second World War. As the term radar indicates, it was used to detect (i) presence of an enemy (aircraft) and (ii) determine its range, i.e., distance from the radar trans-

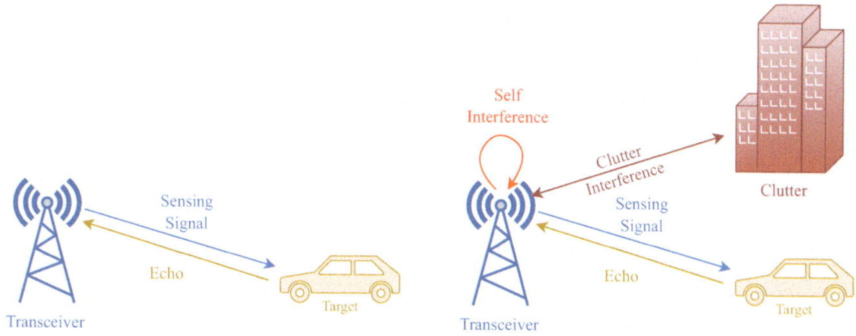

Fig. 4.5 Illustration of wireless sensing scenarios. (Left) represents an overly simplified situation with only target present; (Right) shows the presence of unwanted objects in the environment that reflect their echoes to the sensing transceiver

mitter. Over time different types and implementations of radars have been developed, however, the primary goal is still to determine the range and velocity of the desired target. Consequently, we will look at these two primary sensing parameters, followed by some other (modern) approaches used in wireless sensing.

4.2.4.1 Range Estimation

Time-of-Flight-Based Ranging

The most commonly used method of estimating distance of a target is the using time-of-flight. In this approach the time difference between the instances of signal transmission and reception is measured. Then, leveraging the knowledge of the speed of the radio signal, the distance of the target is determined using the following expression:

$$R = c\tau/2, \tag{4.2}$$

where R denotes the range, c represents the speed of light and τ is the time-of-flight. Note that the factor 2 is used when the sensing transmitter and receiver are co-located (or equidistant from the target). The time-of-flight typically used high-power, very short pulses which form the basis of pulsed radar systems.

Signal Strength Based Ranging

This approach relies on the relation of signal strength attenuation and the distance covered by the radio signal. For instance, according to the log-normal distance pathloss model, the received signal strength $P(d)$ at a distance d is given as [6]

$$P(d) = P(d_0) + 10\gamma \log(d/d_0) + \eta, \tag{4.3}$$

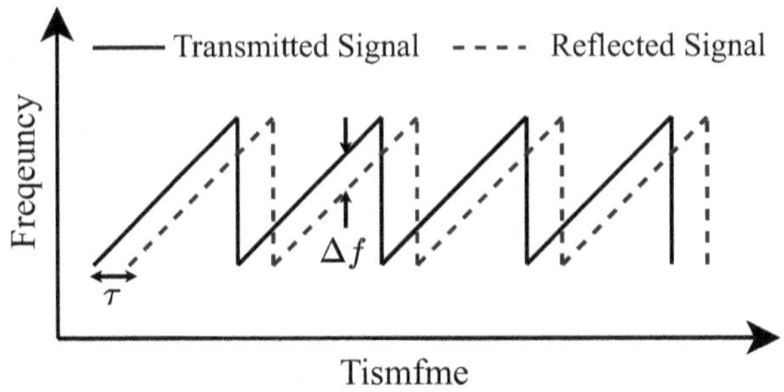

Fig. 4.6 Using FMCW to determine range of the object. The solid line represents the transmitted signal while the dashed line represents the echo signal received from the target. To estimate the range and velocity of the target, time-of-flight τ and beat frequency Δf is calculated

where $P(d_0)$ is the received power at a reference distance d_0, γ is the pathloss exponent and η is the zero-mean Gaussian noise. Using the expression above an estimate of the distance can be obtained. However, note that these equations are typically obtained after curve-fitting over extensive measurement data. As such, they contain environment/scenario-dependent parameters. Consequently, while they are able to give a rough idea about the range, they may be insufficient for high-resolution applications.

Frequency-Modulated Signal-Based Ranging

This approach involves using an frequency-modulated continuous-wave (FMCW) signal to determine the range of the target. As illustrated in Fig. 4.6, an FMCW signal increases in frequency linearly (say a factor k) over the transmission bandwidth and the same signal is retransmitted after every fundamental time period. Once this transmitted signal is reflected from the target, it is received as the dashed line in Fig. 4.6. At this instant, the difference in the frequencies of the transmitted and received signal (also referred to as *beat frequency* and denoted by Δf) can be used to determine the corresponding time-of-flight τ as

$$\tau = \Delta f / k, \tag{4.4}$$

and the distance can be calculated using (4.2).

4.2.4.2 Velocity Estimation

Continuous-Wave-Based Velocity Estimation

This approach uses a continuous, single frequency sinusoidal signal and leverages the Doppler effect to determine the velocity of the target object. The Doppler effect refers to the apparent change in the observed frequency due to mobility of the transmitter/receiver (or the surrounding) objects in accordance with the following expression:

$$f_d = 2v\frac{f_t}{(c - v)}, \tag{4.5}$$

where f_d is the Doppler frequency, i.e., the difference between the transmitted (f_t) and received/observed frequency (f_r), and v is the velocity of the target.

It should be noted that the continuous-wave signal (with a single frequency) is unable to provide information about the range of the target given there is no time reference available. On the other hand, due to periodic transition of the frequency in FMCW, as illustrated in Fig. 4.6, it can be used to determine the velocity as well.

4.2.4.3 Others (Radio Fingerprinting)

In addition to determining some specific target-related parameters such as its velocity and/or range, wireless sensing can be used to extract much more information about that environment and the object within. For instance, **radio fingerprinting** provides an alternate to range/velocity determination for localization of the target. In this approach, a multitude of measurements (signal power, signal phase) is taken for different positions of the object and a database built. Then, when localization is needed, the measurements are compared to the database and the location is determined. However, one of the limitation of this approach is the need for extensive measurements in the testing environment.

It should be noted that this (sub)section provides a very basic overview of wireless sensing. For readers interested in more in-depth technical discussion of the approaches/algorithms used in this domain, other sources may be consulted.

4.2.5 Foundational Principles of Symbiotic Radio

4.2.5.1 Principle of Biosymbiosis

A natural ecosystem comprises two primary components: biotic and abiotic. The biotic components encompass all living organisms, including animals, plants, and bacteria, while abiotic components refer to non-living physical resources that influence the ecosystem, such as water, soil, and minerals. Various biotic components coexist within a biological ecosystem, forming diverse associations while sharing physical resources necessary for survival. In 1879, pathologist Anton de Bary introduced the term "symbiosis," describing interspecific interactions among different organisms [7]. He defined symbiosis as "interspecific associations, or symbioses, a phenomenon in which two different species depend on each other for food, shelter, or protection." Organisms involved in symbiosis are termed symbionts, and these relationships determine how they interact for resource sharing. Symbiosis can be classified as obligatory or facultative based on the degree of dependence between symbionts. Obligatory symbiosis occurs when one or both symbionts cannot survive independently, while facultative symbiosis allows for independent survival alongside interaction. Symbionts may exhibit various relationships, including mutualism, commensalism, and parasitism [8].

4.2.5.2 Symbiotic Relationships

Mutualism
In this symbiotic relationship, the symbionts mutually benefit from their shared resources. A notable example of such a symbiotic relationship is observed between bees and flowers. In this interaction, bees obtain nectar from flowers, which serves as a food source, while, in return, the flowers receive essential pollination services from the bees. This mutualistic exchange exemplifies how both parties enhance their survival and reproductive success through cooperative interaction.

Commensalism
In this symbiotic relationship, one of the symbionts benefits from the interaction while the other experiences neither benefit nor harm. A clear example of this type of symbiotic relationship can be observed between remora fish and sharks. In this interaction, remora fish attach themselves to sharks, gaining transportation and access to leftover food from the shark's meals, while the sharks remain unaffected by the presence of the remora. This relationship illustrates the nature of commensalism, where one species derives advantages without impacting the other.

Parasitism
In this symbiotic relationship, one symbiont benefits by exploiting the resources of another, resulting in harm to the host organism. A prominent example of this type of relationship is observed between ticks and mammals. Ticks feed on the blood of mammals, deriving essential nutrients for their survival. In return, they can transmit various diseases to their hosts, thereby causing significant harm. This interaction exemplifies parasitism, where one organism benefits at the expense of another, highlighting the complex dynamics within ecological relationships.

4.2.5.3 Symbiotic Radio

The current RF ecosystem is rapidly evolving due to the proliferation of wireless applications; however, radio resources, particularly the frequency spectrum, are limited and insufficient to meet the demands of various systems. In such circumstances, different radio systems must share resources to fulfill their individual requirements. Notably, there is a parallel between the RF ecosystem and biological ecosystems, wherein resources and systems can be likened to biotic and abiotic components. Radio systems utilize resources such as frequency, space, and time to perform essential functions, including communication and sensing. This resource-sharing fosters the development of a symbiotic radio system, where multiple radio systems coexist and establish symbiotic relationships for efficient resource utilization.

Depending on the interdependencies of the systems involved, either obligatory or facultative symbiosis can emerge among the radio systems. An example of obligatory symbiosis occurs when one radio system relies entirely on the resources of another for its survival [9]. Conversely, facultative symbiosis arises when two radio systems can share resources while maintaining the capability to survive independently. Finally, the concept of symbiotic radio offers a framework for the coexistence of diverse radio systems that may perform similar functions, such as communication, or different functions, such as sensing and communication.

4.3 Symbiotic Sensing and Backscatter Communication for Sustainable IoT

Symbiotic sensing and backscatter communication (SSBC) refers to a symbiotic radio system design in which sensing and backscatter communication functionalities are integrated to share the same radio resources and establish a mutually beneficial relationship. This integration not only enables optimal use of available spectrum and energy but also the hardware between sensing and backscatter communication system. Through SSBC, the A-IoT device can leverage these functionalities for enhancing its overall performance and operational sustainability. Specifically, backscatter communication significantly reduces power consumption of the device while sensing helps in identifying the location/position of the A-IoT.

4.3.1 Unified Signal Design

A critical functionality of a SSBC system is the ability to provide a unified signal that simultaneously acquires sensing parameters from the environment (e.g., temperature, humidity, and proximity) and the device (e.g., range, velocity, and position) while supporting backscatter communication. By eliminating the necessity for separate transceivers and distinct signal paths for sensing and communication, the unified signal architecture contributes to a lower power consumption profile for A-IoT devices, thereby extending their operational longevity.

The implementation of a unified signal requires the optimal resource allocation, enabling a central controller or intelligent algorithm to dynamically adjust the balance between sensing and backscatter communication performance based on real-time application demands and environmental conditions. For example, SSBC tries to improve the reliability of A-IoT device's data transmission by minimizing the potential for self-interference and co-channel interference while finding the position accurately.

In the unified signal design the transmitter sends one carrier signal which is utilized by the A-IoT for backscattering its data and by the receiver for obtaining the sensing information. Let $x(t)$ be the unmodulated carrier signal transmitted by the transmitter at time t, and A-IoT modulates its information $b(t)$ and reflect the signal with a reflection coefficient α. The signal x is also reflected from the environment and the physical structure of the backscatter device. The composite received signal can be expressed as

$$y(t) = y_s(t) + y_{\mathrm{bc}}(t) + n(t) \,, \tag{4.6}$$

where $y_s(t) = x(t)\delta(t - \tau)e^{j2\pi f_d t}$ is the received echo from the target, and $y_{\mathrm{bc}} = +h_b(t) * (h_f(t) * \alpha b(t)x(t))$ is the signal received from A-IoT.

Target Detection

To detect the target, the receiver performs signal processing techniques in accordance with the particular type of transmission used and the sensing parameters of interest. For instance, as mentioned earlier FMCW signaling can be used to measure both the range and velocity of the target. This can be achieved by first mixing the received (sensing) signal $x(t - \tau)e^{j2\pi f_d t}$, where τ and f_d denote delay and Doppler frequency, respectively, with the simultaneously transmitted FMCW signal. This process generates an intermediate frequency which includes both the range-induced beat frequency (Δf) and the Doppler frequency shift. Afterward, the signal is passed through a low-pass filter to remove any unwanted high-frequency components. Then, following analog-to-digital conversion, two-dimensional (2D) fast Fourier transform (FFT) is applied. The *range FFT* is applied across the samples from a single sweep (i.e., one time period) of the FMCW signal. The peaks in the resultant FFT output correspond to the ranges of the different target object. After performing the range FFT for different sweeps, a second *Doppler FFT* is performed across the same range bin over successive chirps. This second FFT reveals the Doppler frequency shift, which is directly proportional to the target's velocity.

A-IoT Signal Detection

To detect the A-IoT information, the receiver first removes the estimated sensing signal $\tilde{y_s}(t)$ by applying successive-interference cancellation as

$$\hat{y}(t) = y(t) - \tilde{y_s}(t) \,. \tag{4.7}$$

Afterward, an energy detection method is applied on the $\hat{y}(t)$ to detect the A-IoT signal. For instance, when A-IoT uses an OOK modulation the incident signal is reflected only if b is "1$''$, otherwise A-IoT remains silent. In this situation, the energy detector checks either the received signal contains noise or the information signal. The type of detection scheme is non-coherent and simple, however, this method is prone to errors with high noise variations and residual sensing signal interference. To achieve a reliable backscatter communication, the receiver can apply sophisticated coherent detection methods with channel equalization, and advance interference cancellation mechanisms.

4.3.2 Topologies

The SSBC system can be designed based on three different topologies, including monostatic, bistatic, and multistatic as illustrated in Fig. 4.7. These topologies are defined based on the configuration of the transmitter(s) and receiver(s) in the system [10]. Each of these topologies brings distinct characteristics to the SSBC system.

Monostatic SSBC

As illustrated in Fig. 4.7a, the monostatic case refers to the scenario where the transmitter and receiver (referred to as transceiver) are co-located. In this case, both the forward (h_f) and backward (h_b) are the same. A major advantage of monostatic configuration is the inherent knowledge of the transmitted signal. This results in reduced need for synchronization which, in turn, leads to simpler hardware. On the other hand, the reflected received signals is multiple orders of magnitude weaker compared to the transmitted signals leading to self-interference. So, it is necessary for monostatic devices to ensure isolation between the transmitted and received signals.

Bistatic SSBC

The bistatic configuration is illustrated in Fig. 4.7b. Here, the transmitter and receiver are physically separated. Consequently, the forward (h_f) and backward (h_b) channels are separate. While the figure shows a single receiver doing both the sensing and backscatter signal reception, it is possible to have separate receivers for both applications. A major benefit of the bistatic configuration is the lack of self-interference. On the other hand, bistatic configuration requires additional synchronization or reference signaling between the transmitter and receiver for both communication and sensing applications.

Multistatic SSBC

The multistatic configuration, as illustrated in Fig. 4.7c, can be considered a generalization of the bistatic case. Here, it is possible to have multiple transmitters and receivers (even though the illustration shows a single transmitter). Such configurations can leverage the concept of diversity, where multiple copies of same data (or multiple reading of the same sensing parameter) are used to improve the reliability of the system. However, in this case, the synchronization becomes significantly more complicated since, unlike the previous case, there isn't a single reference point (transmitter).

4.4 Application and Use Cases

Symbiotic sensing and backscatter communication, often working hand-in-hand, offer a revolutionary approach to deploying IoT devices, particularly where power constraints, cost, and longevity are critical. Their ability to operate with ultra-low

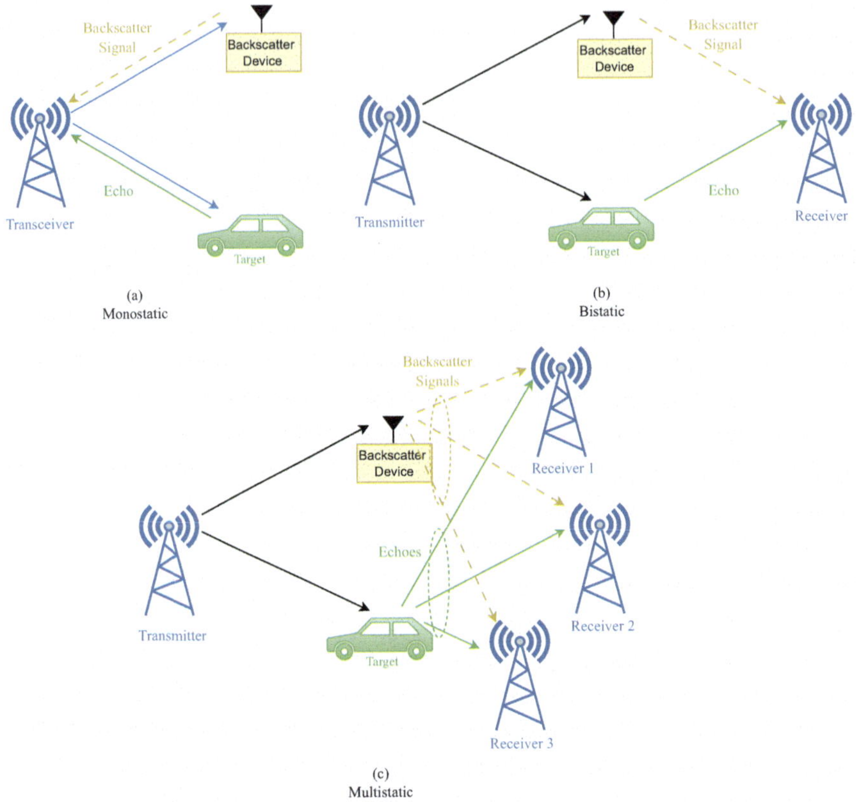

Fig. 4.7 Symbiotic sensing and backscatter communication system Topologies; **a** Monostatic scenario where a single transceiver is used for both sensing and communication operations, **b** Bistatic topology where transmitter and receivers are separate to perform sensing and backscatter communication, **d** Multistatic topology where the more than one receiver is used for sensing and backscatter communication

power by leveraging ambient RF energy makes them ideal for a wide range of practical applications. Some of these have been identified from sensing and backscatter communication perspectives in [11, 12], respectively:

Environmental Observation
One of the main use cases of IoT technologies is environmental monitoring. The use of backscatter or IoT becomes pertinent in this context when one considers the remote deployment scenarios such as dense forests or mountains where these devices may be used for early warning of natural (or man-made) disasters such as forest fires or earthquakes. In such cases, it is not feasible to lay down a dedicated power infrastructure or replace the batteries of the devices repeatedly. Additionally, since the purpose of these signals is to actually sense the environment, the backscatter device is designed in accordance with the sensing signals as the ambient RF source.

Smart Livestock Monitoring
Livestock (or even wildlife) monitoring presents an interesting application of the symbiotic sensing and backscatter communication paradigm. Consider a large number of animals in either a farm or a preservatory park who need to be continuously monitored with periodic readings of their vitals, etc. In such a case, it is not feasible to mount wireless transmitters on the animals but they can be equipped with backscatter devices. Since the animal can wander around in large spaces, it would be much more efficient to first localize or *sense* the animal(s) of interest and then transmit the primary signal in the relevant direction rather than blindly transmitting the signal isotropically. In this application, the sensing is used to improve the backscatter communication performance.

Temperature Monitoring in Smart Packaging of Perishable Goods
Smart packaging offers a vital use for symbiotic sensing and backscatter communication in cold chain monitoring. Imagine delicate goods like vaccines needing constant temperature checks during transport. Traditional bulky loggers are expensive and need batteries changed often.

Here, tiny, battery-free backscatter tags with built-in temperature sensors are placed on each package. They work symbiotically by drawing small amounts of power from existing Wi-Fi or reader signals in warehouses and trucks. This harvested energy allows the sensor to take a temperature reading. The tag then reflects the RF signal back, carrying the temperature data. This ultra-low-power method gives real-time, item-specific temperature tracking, preventing spoilage and cutting waste, all without needing active power on every single package.

Note that in the first use case (*Environment Monitoring*) we highlight the co-design aspect of symbiosis where backscatter communication is to performed on sensing-centric primary signal. Then, the second use case (*Smart Livestock Monitoring*) is representative of “sensing for backscatter communication" approach where the information about the environment obtained from various sources can be used to improve the performance of the communication system. Finally, the last use case (*Temperature Monitoring in Smart Packaging of Perishable Goods*) illustrated the converse, i.e., when backscatter communication is used to monitor or sense certain environmental parameters. Consequently, a wide variety of applications can be developed using the same principles described in the representative use cases.

4.5 Challenges and Future Research Directions

While the integration of symbiotic sensing and backscatter communication offers significant benefits for the sustainable Internet of Things, this advanced paradigm introduces several technical challenges [13]. To fully realize its potential, a thorough analysis of these issues is necessary, ranging from managing complex interference scenarios to developing new performance metrics [14]. The following sections delve into these critical challenges and outline key considerations for future research and system design.

4.5.1 Interference Management

The SSBC system enables sensing and backscatter communication using same radio resources, which effectively enhances the resource sharing and reduce cost of the system. However, the resource sharing brings the challenge of interference that compromises the reliability of the system. The common types of interfere include but not limited to self-interference, mutual interference, clutter/ environmental interference and direct path interference. The self-interference occurs when the transmitter and receiver are collocated such as in the case of monostatic SSBC system. Mutual interference occurs between sensing to backscatter communication signals. Clutter interference occurs mainly due to the unwanted reflectors/objects in the environment that adds up at the receiver along with the desired signal. Finally, the direct path interfere is related to bistatic and multistatic SSBC systems, which occurs from the direct signal (i.e., higher power compared to the sensing and backscattered signals) from the transmitter at the receiver. Mitigating these interference requires advance interference suppression, and mitigation techniques to enhance the reliability of SSBC system, and achieving maximum gains out of it. One possible solution to suppress the interference is to develop SSBC system with the commensal and mutualistic relationships between sensing and backscatter communication.

4.5.2 Signal Design and Processing

The SSBC system requires a signal design that allows both easy extraction of the sensing parameters (e.g., high range resolution, accurate velocity estimation) and backscatter communication performance (e.g., high reliability, data rate). simple modulation (and demodulation) for the backscatter device. However, the sensing requirements (e.g., large bandwidth, short pulses) can contradict with the communication requirements (e.g., robustness to fading, spectral efficiency). Furthermore, backscatter modulation can cause the interference to the sensing signal which may reduce the sensing performance. Additionally, from the backscatter domain this problem is further exacerbated due to the power limitations and cascaded propagation channel. To overcome these challenge sophisticated modulation, demodulation, and receiver design need to be designed.

4.5.3 New Key Performance Indicators

The SSBC system facilitates both sensing and backscatter communication through various symbiotic relationships. While evaluating the performance of each individual system is straightforward and can be conducted using established key performance indicators, such as the probability of miss detection for sensing and bit error rate for backscatter communication, assessing the overall performance of the SSBC system necessitates the development of new key performance indicators. These indicators must accurately reflect the system's performance within the shared resources paradigm.

4.5.4 Resource Allocation

The distribution of resources between sensing and (backstatter) communication represents an interesting optimization challenge. While relevant in all symbiotic situations, this issue might be critical in case of a parasitic relationship. Power allocation and spectral resource allocation needs to be done carefully to meet both the sensing and backscatter communication requirements as mentioned earlier.

4.5.5 System Design

System design is significantly influenced by the chosen topology. For example, bi/multistatic configurations can provide numerous advantages; however, they necessitate synchronization among nodes, which becomes increasingly complex when both communication and sensing services are integrated. Therefore, selecting the appropriate topology that ensures optimal sensing and communication performance requires a thorough analysis of these configurations under dynamic SSBC systems.

4.6 Summary

This chapter comprehensively explores symbiotic radio, integrating sensing and backscatter communication for sustainable A-IoT applications. It begins by establishing the foundational principles of symbiotic radio, emphasizing its role in promoting energy efficiency and sustainability by enabling low-power devices to reuse ambient RF signals. Key 3GPP use cases, including sensor data collection and inventory management, are discussed. The chapter analyzes communication topologies like monostatic and bistatic configurations, providing examples where sensing improves backscatter communication performance, such as localizing mobile targets to enhance data transmission. Furthermore, it addresses unified signal and system design considerations for simultaneous sensing and backscatter communication. The conclusion highlights potential applications, challenges like primary signal design and resource allocation, and future research directions for these evolving A-IoT technologies.

Acknowledgements This work is supported in part by The Scientific and Technological Research Council of Türkiye (TÜBİTAK) 1515 Frontier R&D Laboratories Support Program for Türk Telekom 6G R&D Lab under project number 5249902 and in part by the Kocaeli University Scientific Research Projects Division (BAP) Joint Collaboration Project Program under grant number 3760.

References

1. 6G Research Visions: White Paper on Critical and Massive Machine Type Communication Towards 6G. University of Oulu, Oulu, Finland (2020)
2. Sinha, S.: State of IoT 2024: Number of connected IoT devices growing 13% to 18.8 billion globally (2024). https://iot-analytics.com/number-connected-iotdevices/
3. UN Department of Economic and Social Affairs: The 17 Goals. https://sdgs.un.org/goals
4. Butt, M.M., Mangalvedhe, N.R., Pratas, N.K., Harrebek, J., Kimionis, J., Tayyab, M., Barbu, O.-E., Ratasuk, R., Vejlgaard, B.: Ambient IoT: a missing link in 3GPP IoT devices landscape. IEEE Internet Thing J. **7**(2), 85–92 (2024)
5. Stockman, H.: Communication by means of reflected power. Proc. IRE **36**(10), 1196–1204 (1948)
6. Seidel, S.Y., Rappaport, T.S.: 914 MHz path loss prediction models for indoor wireless communications in multifloored buildings. IEEE Trans. Antennas Propag. **40**(2), 207–216 (1992)
7. Martin, B.D.S.E.: SymbiosisâŁ¯: "Living Together" in Chaos. Studies in the History of Biology, vol. 4, no. 4, pp. 7–25 (2012)
8. Montesinos, E.: Plant-Associated Microorganisms: A View from the Scope of Microbiology. Springer (2003)
9. Kuschel, H., Cristallini, D., Olsen, K.E.: Tutorial: passive radar tutorial. IEEE Aerosp. Electron. Syst. Mag. **34**(2), 2–19 (2019). https://doi.org/10.1109/MAES.2018.160146
10. 3rd Generation Partnership Project: 3GPP TR 38.848: Study on Ambient IoT (Internet of Things) in RAN (Release 18) V18.0.0 (2023)
11. 3rd Generation Partnership Project: 3GPP TR 22.837: Feasibility Study on Integrated Sensing and Communication (Release 19) V19.4.0 (2024)
12. 3rd Generation Partnership Project: 3GPP TR 22.840: Study on Ambient powerenabled Internet of Things (Release 19) V19.0.0 (2024)
13. Zheng, K., Xu, R., Mei, J., Yang, H., Lei, L., Wang, X.: Ambient IoT towards 6G: standardization, potentials, and challenges. IEEE Access **12**, 146668–146677 (2024)
14. 3rd Generation Partnership Project: 3GPP TR 38.769: Study on solutions for Ambient IoT (Internet of Things) in NR (Release 19) V19.0.0 (2024)

Chapter 5
Adaptive Nonlinear State Estimators Based on RKHS for Future Applications in Integrated Sensing and Communication

Uday Kumar Singh, Rangeet Mitra, Vimal Bhatia, Amit Kumar Mishra, and K. Venkateswaran

Abstract Integrated Sensing and Communication (ISAC) systems are poised to play a pivotal role in the evolution of next-generation wireless networks by merging communication and sensing functionalities into a unified framework. ISAC enables a wide range of applications, such as autonomous vehicles, smart cities, and IoT networks, by offering precise real-time state estimation from various sensor data. One of the key challenges in ISAC is estimating system states, such as position, velocity, and acceleration, from noisy, uncertain, and nonlinear sensor measurements. Traditional state estimation techniques, such as Kalman filtering (KF), particle filtering, and their nonlinear variants, depend on an accurate measurement model, which is often difficult to derive in dynamic and heterogeneous environments. These methods also typically assume Gaussian distributions for sensor data, a condition that may not hold in real-world scenarios. To address these challenges, this chapter explores the use of Reproducing Kernel Hilbert Space (RKHS) techniques, which allow for direct estimation of the measurement model from data. This eliminates the need for precise prior knowledge about the model or its probability distribution. The RKHS-estimated model is then integrated with conventional Bayesian state estimation techniques,

U. K. Singh (✉)
Department of ECE, SR University, Warangal, Telangana, India
e-mail: udaykumar.singh@sru.edu.in

R. Mitra · K. Venkateswaran
Department of ECE, CMRIT, Bengaluru, Karnataka, India
e-mail: rangeet.m@cmrit.ac.in

K. Venkateswaran
e-mail: venkateswaran.k@cmrit.ac.in

V. Bhatia
Discipline of Electrical Engineering IIT Indore, Indore, India
e-mail: vbhatia@iiti.ac.in

A. K. Mishra
Faculty of Space Technologies, AGH University of Kraków, Kraków, Poland
e-mail: akmishra@ieee.org

A. K. Mishra and V. Bhatia (eds.), *Integrated Sensing and Communication for Sustainable Development Goals*, Signals and Communication Technology,
https://doi.org/10.1007/978-981-92-0209-6_5

creating a probabilistic framework that updates state estimates while accounting for both measurement uncertainty and noise. Through practical case studies, the chapter illustrates how combining RKHS-based measurement models with Bayesian filters improves the accuracy and robustness of state estimation, even when dealing with highly nonlinear and non-Gaussian sensor data.

Keywords RKHS · ISAC · IoT · Bayesian filters · EKF

5.1 Introduction

This chapter explores the transformative potential of Integrated Sensing and Communication (ISAC) technology in the realization of autonomous devices by combining advanced sensor signal processing with communication capabilities. ISAC is essential for applications such as autonomous vehicle navigation, vehicle cabin monitoring, and wireless vital signs monitoring, facilitating the seamless integration of IoT devices. This integration enables real-time data collection and communication, allowing devices to make informed decisions that improve safety, efficiency, and convenience. A key element in these applications is the accurate estimation of dynamic states, including location, velocity, acceleration, and sometimes the shape of surrounding objects. Such estimations are crucial for ISAC systems to respond effectively to new and unknown situations, ensuring the safety and security of the host [1, 2, 5, 9, 12, 18].

The chapter investigates state estimation problems, addressing long-standing challenges, and offering effective solutions. It begins by discussing the limitations of traditional state estimation techniques, such as the Kalman filter (KF), which performs well for linear systems but faces difficulties with nonlinear sensor measurements and models. For example, sensors that provide measurements in spherical coordinates cannot be directly modeled using linear approaches. This challenge necessitates the use of nonlinear extensions like the Extended Kalman Filter (EKF) or Unscented Kalman Filter (UKF). However, these methods assume Gaussian noise distributions, an assumption that often does not hold in real-world applications, where noise and disturbances may deviate from Gaussian behavior. To address these limitations, the chapter introduces recent advancements in adaptive algorithms based on Reproducing Kernel Hilbert Space (RKHS) techniques [13–15]. The evolution of these algorithms is traced, emphasizing their ability to estimate kinematic states in the face of uncertainties in measurement models and noise distributions. Practical case studies are provided to showcase the effectiveness and robustness of RKHS-based methods in overcoming the shortcomings of traditional Bayesian filters in nonlinear and non-Gaussian environments.

The chapter begins by introducing the core concepts of the state estimation problem, offering a comprehensive understanding of the challenges at hand[1]. It then

[1] This chapter looks at radar sensors. The ideas presented could be expanded to other common IoT sensors, such as LIDAR and ultrasonic sensors.

explores three distinct adaptive modifications to the Bayesian filter, each crafted to enhance its robustness against the challenges of nonlinearity and non-Gaussianity encountered in real-world scenarios. Each section concludes with a demonstration of the practical effectiveness of these adaptive algorithms and their improved performance in complex environments. The chapter ends with a summary of the key findings and conclusions drawn from the case studies.

Notations: Scalar variables (constants) are denoted by lower (upper) case letters. Vectors (matrices) are denoted by boldface lower (upper) case letters. The superscripts $(.)^T$, $(.)^H$, and $(.)^*$ denote matrix/vector transpose, complex conjugate transpose, and complex scalar conjugation operation, respectively. $\mathbb{E}[.]$ denotes statistical expectations and $\mathbb{R}$ denotes the set of real numbers. $\mathbf{I}_n$ denotes the cardinality identity matrix n and $\otimes$ denotes the Kronecker product.

5.2 Problem Statement

Let the evolution of the target state vector with cardinality n_s ($\mathbf{s}_k \in \mathbb{R}^{n_s \times 1}$) at any arbitrary time instance k be given by

$$\mathbf{s}_k = \mathbf{f}(\mathbf{s}_{k-1}) + \mathbf{q}_k, \tag{5.1}$$

where $\mathbf{s}_{k-1}$ is the previous target state vector, $\mathbf{f} \in \mathbb{R}^{n_s \times 1}$ governs the motion of the target, and $\mathbf{q}_k \in \mathbb{R}^{n_s \times 1}$ accounts for uncertainties in the modeling of the state model.

Corresponding to (5.1), the sensor measurements with cardinality n_m at kth time instance ($\mathbf{y}_k \in \mathbb{R}^{n_m \times 1}$) are given by

$$\mathbf{y}_k = \mathbf{h}(\mathbf{s}_k) + \mathbf{w}_k, \tag{5.2}$$

where $\mathbf{h} \in \mathbb{R}$ is the nonlinearity introduced by the sensor and $\mathbf{w}_k \in \mathbb{R}^{n_m \times 1}$ accounts for the error in estimating $\mathbf{y}_k$ using the sensor.

In (5.1), the components of $\mathbf{s}_k$ depend on the type of target motion model. In the realm of state estimation problems, three different practical target motion models are possible. The associated $\mathbf{f}(\mathbf{s}_{k-1})$ for these three different target motion models are given by [2]:

1. *Constant displacement*: The target state vector is composed of the varying location of the target in the 3 D space, i.e., $\mathbf{s}_k = [x_k, y_k, z_k]^T$ and at each k time instance $\mathbf{s}_k$ it is increased by the constant factor Δs. Consequently, the noise-free $\mathbf{f}(\mathbf{s}_{k-1})$ is given by

$$\mathbf{f}(\mathbf{s}_{k-1}) = \mathbf{s}_{k-1} + \Delta s. \tag{5.3}$$

2. *Constant velocity*: In this classical setting, the target state vector is composed of dynamic target location in 3D space and the corresponding constant velocities

(v_{xk}, v_{yk}, v_{zk}) along the x-, y-, and z-axes, i.e., $\mathbf{s}_k = [x_k, v_{xk}, y_k, v_{yk}, z_k, v_{zk}]^T$. Hence, the noise-free $\mathbf{f}(\mathbf{s}_{k-1})$ is given by

$$\mathbf{f}(\mathbf{s}_{k-1}) = \begin{bmatrix} 1 & T_{\text{cpi}} & 0 & 0 & 0 & 0 \\ 0 & 1 & 0 & 0 & 0 & 0 \\ 0 & 0 & 1 & T_{\text{cpi}} & 0 & 0 \\ 0 & 0 & 0 & 1 & 0 & 0 \\ 0 & 0 & 0 & 0 & 1 & T_{\text{cpi}} \\ 0 & 0 & 0 & 0 & 0 & 1 \end{bmatrix} \mathbf{s}_{k-1}, \tag{5.4}$$

where T_{cpi} is the coherent pulse interval (CPI) of the sensor. It should be worth mentioning that the constant velocity (CV) and constant displacement (CD) models evolve the target locations similarly. However, the two models differ in the content of $\mathbf{s}_k$ and the CV model gives more information about the target's states.

3. *Constant acceleration*: In this refined model, the target state vector is composed of target location in the 3 D space, corresponding varying velocities (v_{xk}, v_{yk}, v_{zk}) and constant accelerations (a_{xk}, a_{yk}, a_{zk}) along the x-, y-, and z-axes, that is, $\mathbf{s}_k = [x_k, v_{xk}, a_{xk}, y_k, v_{yk}, a_{yk}, z_k, v_{zk}, a_{zk}]^T$. Therefore, the noise-free $\mathbf{f}(\mathbf{s}_{k-1})$ is given by

$$\mathbf{f}(\mathbf{s}_{k-1}) = \begin{bmatrix} 1 & T_{\text{cpi}} & 0.5T_{\text{cpi}}^2 & 0 & 0 & 0 & 0 & 0 & 0 \\ 0 & 1 & T_{\text{cpi}} & 0 & 0 & 0 & 0 & 0 & 0 \\ 0 & 0 & 1 & 0 & 0 & 0 & 0 & 0 & 0 \\ 0 & 0 & 0 & 1 & T_{\text{cpi}} & 0.5T_{\text{cpi}}^2 & 0 & 0 & 0 \\ 0 & 0 & 0 & 0 & 1 & T_{\text{cpi}} & 0 & 0 & 0 \\ 0 & 0 & 0 & 0 & 0 & 1 & 0 & 0 & 0 \\ 0 & 0 & 0 & 0 & 0 & 0 & 1 & T_{\text{cpi}} & 0.5T_{\text{cpi}}^2 \\ 0 & 0 & 0 & 0 & 0 & 0 & 0 & 1 & T_{\text{cpi}} \\ 0 & 0 & 0 & 0 & 0 & 0 & 0 & 0 & 1 \end{bmatrix} \mathbf{s}_{k-1}.$$

Using Eqs. (5.1) and (5.2), the primary goal of the state estimation problem is to accurately and effectively estimate $\mathbf{s}_k$ from $\mathbf{y}_k$. Traditionally, depending on the characteristics of $\mathbf{h}$, KF, EKF, or UKF is used. Specifically, if $\mathbf{h}$ is linear, the KF is sufficient; however, if $\mathbf{h}$ is nonlinear, the EKF and UKF are more suitable and produce satisfactory results. In the EKF, $\mathbf{h}$ is approximated using its Jacobian matrix evaluated at the predicted state vector. Similarly, the UKF approximates $\mathbf{h}$ using unscented sigma points. Both EKF and UKF rely on nonlinear measurement function approximations, and this presents an opportunity to improve their performance by replacing these approximations with a more accurate estimate of $\mathbf{h}$, denoted as $\hat{\mathbf{h}}$. Moreover, in many practical applications, the exact form of $\mathbf{h}$ is not known. For instance, sensors may not always provide measurements in the same format, making the Jacobian derived for one sensor unsuitable for another. Furthermore, $\mathbf{w}_k$ does not necessarily follow a Gaussian distribution, since measurement noise is essentially estimation noise, which cannot always be assumed to be Gaussian. Furthermore,

Table 5.1 RKHS-based adaptive variant of EKF in accordance with the nature of $\mathbf{w}_k$ and $\mathbf{h}$

$\mathbf{w}_k$	$\mathbf{h}$	EKF adaptive variant	Criterion
Gaussian	Nonlinear	EKF-RKHS	MMSE
Non-Gaussian	Nonlinear	EKF-MCC-RKHS	MCC
Non-Gaussian	Nonlinear	EKF-KRMEE	MEE

various environmental effects around the sensor, along with random target movements, challenge the traditional Gaussian assumption for measurements.

In this chapter, the challenge of dealing with nonlinearity and non-Gaussianity in the context of state estimation is addressed through three adaptive variants of the EKF based on RKHS: EKF-RKHS, EKF-MCC-RKHS, and EKF-KRMEE. EKF-RKHS employs the minimum mean square error (MMSE) approach, which is effective when dealing with Gaussian noise. However, EKF-MCC-RKHS and EKF-KRMEE incorporate the maximum correntropy criterion (MCC) and minimum error entropy (MEE) methods, respectively, which enhance performance in the presence of non-Gaussian noise. In all these adaptive filters, the vector $\mathbf{h}$ is estimated by optimizing the corresponding cost function and subsequently used in the EKF to perform the usual prediction and update steps. A brief classification of these three adaptive filters, based on the characteristics of $\mathbf{w}_k$ and $\mathbf{h}$, is provided in Table 5.1, with detailed explanations offered in their respective sections.

5.3 Adaptive EKF Utilizing MMSE

This section provides a detailed description of using EKF-RKHS to estimate $\mathbf{s}_k$. It begins with a brief overview of the standard EKF, followed by an explanation of how the EKF is modified to create EKF-RKHS.

5.3.1 EKF

For the state and measurement model described, respectively, in (5.1) and (5.2), let at kth time instance the available estimate of $\mathbf{s}_{k-1}$ be $\hat{\mathbf{s}}_{k-1}$ and the corresponding error covariance matrix be $\mathbf{P}_{k-1} = \mathbb{E}[(\mathbf{s}_{k-1} - \hat{\mathbf{s}}_{k-1})(\mathbf{s}_{k-1} - \hat{\mathbf{s}}_{k-1})^T]$. The EKF propagates $\hat{\mathbf{s}}_{k-1}$ and $\mathbf{P}_{k-1}$ iteratively and produces the final estimate of $\mathbf{s}_k$ in two steps (a) prediction and (b) update. In the prediction step, the EKF predicts $\mathbf{s}_k$ and $\mathbf{P}_k$ with the help of the following equations:

$$
\begin{aligned}
\mathbf{s}_k^p &= \mathbf{F}_{k-1}\hat{\mathbf{s}}_{k-1}, \\
\mathbf{P}_k^p &= \mathbf{F}_{k-1}\mathbf{P}_{k-1}\mathbf{F}_{k-1}^T + \mathbf{Q}_s,
\end{aligned}
\tag{5.5}
$$

where $\mathbf{F}_{k-1} \in \mathbb{R}^{n_s \times n_s}$ is the motion model matrix and can be obtained, respectively, from (5.3), (5.4), or (5.2), depending on the motion model, and $\mathbf{Q}_s \in \mathbb{R}^{n_s \times n_s}$ is the covariance matrix of $\mathbf{q}_k$.

After prediction, in the update step, the predictions made in (5.5) are updated with the help of the available sensor measurement ($\mathbf{y}_k$). The following equations give the update equations of EKF:

$$\begin{aligned} \mathbf{y}_k^p &= \mathbf{H}_k \mathbf{s}_k^p, \\ \mathbf{K}_k &= \mathbf{P}_k^p \mathbf{H}_k^T (\mathbf{H}_k \mathbf{P}_k^p \mathbf{H}_k^T + \mathbf{R}_m)^{-1}, \\ \hat{\mathbf{s}}_k &= \mathbf{s}_k^p + \mathbf{K}_k (\mathbf{y}_k - \mathbf{y}_k^p), \\ \mathbf{P}_k &= \mathbf{P}_k^p - \mathbf{K}_k \mathbf{H}_k \mathbf{P}_k^p, \end{aligned} \tag{5.6}$$

where $\mathbf{H}_k = \left.\frac{\partial \mathbf{h}(\mathbf{s}_k)}{\partial \mathbf{s}_k}\right|_{\mathbf{s}_k = \mathbf{s}_k^p} \in \mathbb{R}^{n_m \times n_s}$ is the Jacobain matrix of $\mathbf{h}$ (evaluated at prediction $\mathbf{s}_k^p$), $\mathbf{y}_k^p$ is the predicted measurement, $\mathbf{K}_k$ is the Kalman gain, and $\hat{\mathbf{s}}_k$ and $\mathbf{P}_k$ are the final estimate of $\mathbf{s}_k$ and the updated error coefficient matrix, respectively.

From (5.6), it is explicitly shown that EKF linearizes $\mathbf{h}$ via Jacobian, which yields the first-order approximation of $\mathbf{h}$. Consequently, it restricts EKF from achieving high estimation accuracy and, in some cases, from yielding inaccurate results. Also, to obtain the Jacobian and, hence, to implement the next steps, EKF must know the exact form of $\mathbf{h}$ (which depends on the sensor type and, in most cases, is not known a priori). In the following subsection, an approach to deal with these shortcomings in EKF is presented. The implementation of RKHS-based EKF (EKF-RKHS) is described, which first estimates $\mathbf{h}$ in RKHS using a well-known RKHS algorithm (KRLS). Subsequently, the estimate of $\mathbf{h}$ is plugged into EKF to perform a further prediction and update.

The advantages of using EKF-RKHS over EKF are twofold: (1) Unlike EKF, the implementation of EKF-RKHS is not restricted to knowing the exact form of $\mathbf{h}$. (2) Since, unlike EKF, the estimate of $\mathbf{h}$ is used, EKF-RKHS will produce the estimate of $\mathbf{s}_k$ with higher precision.

5.3.2 EKF-RKHS

In this subsection, the description of the KRLS algorithm used in the implementation of EKF-RKHS is provided. Firstly, referring to (5.2), the hidden state $\mathbf{s}_k$ is mapped to a high-dimensional RKHS ($\mathcal{H}$) through an unknown implicit mapping function $\phi(\cdot)$. Subsequently, the estimate of $\mathbf{y}_k$ in $\mathcal{H}$ is given by the well-known Representer's theorem [3, 10, 11] as

$$\hat{\mathbf{y}}_k = \langle \boldsymbol{\omega}, \phi(\mathbf{s}_k) \rangle_{\mathcal{H}}, \tag{5.7}$$

where $\langle(\cdot,\cdot)\rangle_{\mathcal{H}}$ is the inner product in $\mathcal{H}$ and $\boldsymbol{\omega}$ is the unknown weight vector in $\mathcal{H}$.

In the search for the optimized value of $\boldsymbol{\omega}$ and for the available pair of $\mathbf{y}_k$ and $\mathbf{s}_k$, $\{(\mathbf{s}_0,\mathbf{y}_0),\ldots,(\mathbf{s}_{k-1},\mathbf{y}_{k-1})\}$, the KRLS algorithm minimizes the following cost function:

$$\mathcal{J}(\boldsymbol{\omega}) = \min_{\boldsymbol{\omega}} \sum_{j=0}^{k-1} |\mathbf{y}_j - \hat{\mathbf{y}}_j|^2 + \lambda\|\boldsymbol{\omega}\|^2, \tag{5.8}$$

where λ is the regularization factor.

Utilizing (5.7) yields

$$\mathcal{J}(\boldsymbol{\omega}) = \min_{\boldsymbol{\omega}} \sum_{j=0}^{k-1} |\mathbf{y}_j - \langle\boldsymbol{\omega},\phi(\mathbf{s}_j)\rangle_{\mathcal{H}}|^2 + \lambda\|\boldsymbol{\omega}\|^2. \tag{5.9}$$

The detailed steps of solving (5.9) are given in [4]. After solving (5.9), $\boldsymbol{\omega}$ is given by

$$\boldsymbol{\omega} = \sum_{j=0}^{k-1} a_j\phi(\mathbf{s}_j) = \mathbf{a}_{k-1}\boldsymbol{\Phi}_{k-1}, \tag{5.10}$$

where $\mathbf{a}_{k-1} = [a_0,\ldots,a_{k-1}]$ and $\boldsymbol{\Phi}_{k-1} = [\phi(\mathbf{s}_0),\ldots,\phi(\mathbf{s}_{k-1})]^T$.

Substituting (5.10) in (5.7) yields

$$\hat{\mathbf{y}}_k = [\langle\phi(\mathbf{s}_0),\phi(\mathbf{s}_k)\rangle_{\mathcal{H}},\ldots,\langle\phi(\mathbf{s}_{k-1}),\phi(\mathbf{s}_k)\rangle_{\mathcal{H}}]\mathbf{a}_{k-1}^T. \tag{5.11}$$

From (5.11), it can be readily analyzed that the estimate of $\mathbf{h}$ at kth time instant is given by

$$\hat{\mathbf{h}}_k = [\langle\phi(\mathbf{s}_0),\phi(\mathbf{s}_k)\rangle_{\mathcal{H}},\ldots,\langle\phi(\mathbf{s}_{k-1}),\phi(\mathbf{s}_k)\rangle_{\mathcal{H}}]. \tag{5.12}$$

For simplifying (5.11), one can use Mercer's theorem [7], which states that the inner product between any two functions $\{\phi(\mathbf{s}_i),\phi(\mathbf{s}_j)\}$ in $\mathcal{H}$ can be equivalently evaluated in Euclidean space by using a reproducing kernel $\kappa(\cdot,\cdot)$ as

$$\kappa(\mathbf{s}_i,\mathbf{s}_j) = \langle\phi(\mathbf{s}_i),\phi(\mathbf{s}_j)\rangle_{\mathcal{H}}.$$

Hence, utilizing Mercer's theorem, the simplified form of $\hat{\mathbf{h}}_k$ is given by

$$\hat{\mathbf{h}}_k = [\kappa(\mathbf{s}_0,\mathbf{s}_k),\ldots,\kappa(\mathbf{s}_{k-1},\mathbf{s}_k)]. \tag{5.13}$$

From (5.13), inference can be drawn that $\mathbf{h}$ is iteratively estimated, and at kth instant, the estimate is based on the current state vector $\mathbf{s}_k$ and the past state vectors $\{\mathbf{s}_0,\ldots,\mathbf{s}_{k-1}\}$. This implies that at kth instant after obtaining $\hat{\mathbf{s}}_k$ from EKF, in (5.35), $\hat{\mathbf{s}}_k$ replaces $\mathbf{s}_k$. Afterward, $\hat{\mathbf{h}}_k$ replaces $\mathbf{h}$ in EKF to perform the further prediction and update steps. Henceforth, the process repeats iteratively for $k = 1, 2, \ldots, K$. The pseudo-algorithm for EKF-RKHS is given in Algorithm 1, where $\mathbf{z}_k$, r_k, $\mathbf{Q}_k$, $\mathbf{e}_k$, and $\mathbf{a}_k$ are according to [4].

Algorithm 1: Implementation of EKF-RKHS

Intialization :

$\mathbf{s}_0, \mathbf{y}(\mathbf{s}_0), \hat{\mathbf{s}}_0, \mathbf{P}_0, \hat{\mathbf{h}}_0 = [\kappa(\hat{\mathbf{s}}_0, \hat{\mathbf{s}}_0)], \mathbf{Q}_0 = (\lambda + \kappa(\hat{\mathbf{s}}_0, \hat{\mathbf{s}}_0))^{-1}, \mathbf{a}_0 = [1, \ldots, 1]^{1 \times n_m}$

for $k = 1, 2, 3, \cdots, K$ **do**

EKF

Compute $\mathbf{s}_k^p$ and $\mathbf{P}_k^p$ using (5.5)

$\mathbf{y}_k^p = \mathbf{a}_{k-1}^T \hat{\mathbf{h}}_{k-1}^T$

$\hat{\mathbf{H}}_k = \left\{ \left(\frac{\partial \hat{\mathbf{h}}_{k-1}}{\partial \hat{\mathbf{s}}_{k-1}} \bigg|_{\hat{\mathbf{s}}_{k-1} = \mathbf{s}_k^p} \right) \mathbf{a}_{k-1} \right\}^T$

Compute $\mathbf{K}_k$, $\hat{\mathbf{s}}_k$, and $\mathbf{P}_k$ using (5.6)

KRLS

$\mathbf{z}_k = \mathbf{Q}_{k-1} \hat{\mathbf{h}}_{k-1}^T$

$r_k = \lambda + \kappa(\hat{\mathbf{s}}_k, \hat{\mathbf{s}}_k) - \mathbf{z}_k^T \hat{\mathbf{h}}_{k-1}^T$

$\mathbf{Q}_k = r_k^{-1} \begin{bmatrix} \mathbf{Q}_{k-1} r_k + \mathbf{z}_k \mathbf{z}_k^T & -\mathbf{z}_k \\ -\mathbf{z}_k^T & 1 \end{bmatrix}$

$\mathbf{e}_k = \mathbf{y}_k - \mathbf{a}_{k-1}^T \hat{\mathbf{h}}_{k-1}^T$

$\mathbf{a}_k = \begin{bmatrix} \mathbf{a}_{k-1} - r_k^{-1} \mathbf{z}_k \mathbf{e}_k^T \\ r_k^{-1} \mathbf{e}_k^T \end{bmatrix}$

$\hat{\mathbf{h}}_k = [\kappa(\hat{\mathbf{s}}_0, \hat{\mathbf{s}}_k), \ldots, \kappa(\hat{\mathbf{s}}_{k-1}, \hat{\mathbf{s}}_k)]$

end

5.3.3 Simulation Results and Inference

In this section, a detailed analysis of the simulation results performed on the target motion models described in Sect. 5.2 is provided. The performance of both EKF and EKF-RKHS is simulated for $K = 200$ time instances. The three target motion models, constant displacement, constant velocity, and constant acceleration, are, respectively, termed the scenario I, scenario II, and scenario III. For all scenarios $T_{\text{cpi}} = 10^{-3}$ s. The theoretical mean square error (TMSE) is used as a metric to compare the performance of EKF and EKF-RKHS. The TMSEs corresponding to the elements of $\hat{\mathbf{s}}_k$ are defined by

$$\text{TMSE} = \mathbb{E}[(\mathbf{s}_k(i) - \hat{\mathbf{s}}_k(i))^2]; \quad i = 1, 2, \ldots, n_s$$

where i denotes the ith element of $\mathbf{s}_k$ and $\hat{\mathbf{s}}_k$. TMSE(i) is obtained by reading the diagonal elements of updated $\mathbf{P}_k$.

For all scenarios, $\mathbf{y}_k$ is assumed to be available in spherical coordinate (as is the case in practice), that is, $\mathbf{y}_k = [r_k, \theta_k, \phi_k]^T$. Consequently, $\mathbf{h}(\mathbf{s}_k)$ is given by

$$\mathbf{h}(\mathbf{s}_k) = \left[\sqrt{x_k^2 + y_k^2 + z_k^2}, \tan^{-1}\left(\frac{y_k}{x_k}\right), \tan^{-1}\left(\frac{\sqrt{x_k^2 + y_k^2}}{z_k}\right) \right]^T.$$

Furthermore, for all scenarios, it is assumed that $\mathbf{Q}_s = \sigma_s^2 \mathbf{I}_{n_s}$ and $\mathbf{R}_m = \sigma_m^2 \mathbf{I}_{n_m}$, where σ_s and σ_m are chosen 0.01 and 0.25, respectively. For EKF-RKHS, the most commonly used Gaussian kernel with kernel width (σ) is used, that is, $\kappa(\mathbf{s}_i, \mathbf{s}_j) = \exp\left(- \frac{\|\mathbf{s}_i - \mathbf{s}_j\|^2}{2\sigma^2} \right)$. In simulations, λ and σ are chosen to obtain the optimum performance of EKF-RKHS. For all scenarios, λ and σ are fixed at 4×10^{-3} and 6, respectively. To simulate EKF and EKF-RKHS for the indoor environment, it is assumed that the target moves in the 3 D space along the x-, y-, and z-axes with a maximum distance of 10 m. Also, for scenario II and scenario III, the maximum values of the target velocities v_{xk}, v_{y_k}, and v_{zk} are limited to 2 m/sec and a_{xk}, a_{y_k}, and a_{zk} are chosen from the bounded set $[-2, 2]$ m/sec^2. In addition, in the following subsections, the simulation setup and the performance analysis of both EKF and EKF-RKHS for all scenarios are discussed in detail.

Constant displacement (scenario I): For Scenario I, initial value of $\mathbf{s}_k$, that is, $\mathbf{s}_0$ is chosen as $\mathbf{s}_0 = [x_0, y_0, z_0]^T$ and $\mathbf{s}_k$ is obtained by advancing $\mathbf{s}_{k-1}$ as in (5.1) with $\Delta\mathbf{s} = [0.01\text{m}, 0.01\text{m}, 0.01\text{m}]^T$. Estimate $\hat{\mathbf{s}}_k$ is obtained at each kth time instance using EKF and EKF-RKHS. For both EKF and EKF-RMSE, the TMSEs are obtained along the x-, y-, and z-axes.

The estimated trajectory using EKF and EKF-RKHS, the corresponding true trajectory (Ground truth), and measurements in the rectangular coordinate are shown in Fig. 5.1. As shown in Fig. 5.1, the estimated trajectory obtained from EKF-RKHS is closer to the ground truth compared to the trajectory estimated from EKF. The improved performance of EKF-RKHS is further shown in Fig. 5.2, which shows the TMSEs in the estimation of location in the x-, y-, and z-axes. In Fig. 5.2, it is observed that the TMSEs for EKF-RKHS are lower than the TMSEs for EKF and that the TMSEs for EKF-RKHS and EKF converge at nearly identical time instances.

Constant velocity (scenario II): Here,

$$\mathbf{s}_0 = [1\text{m}, 2\text{m/s}, 1\text{m}, 1\text{m/s}, 1\text{m}, 0.5\text{m/s}]^T$$

and $\mathbf{s}_k$ is obtained from (5.2).

As shown in Fig. 5.3, the EKF-RKHS yields the estimated trajectory closer to the ground truth. Furthermore, Figs. 5.4 and 5.5 depict, respectively, the TMSEs for the estimated locations and velocities along the x-, y-, and z-axes. From Figs. 5.4 and 5.5, it is explicitly shown that, unlike scenario I, the TMSEs for EKF-RKHS are converging slower. However, for both locations and velocities, EKF-RKHS achieves lower TMSEs in the x-, y-, and z-directions compared to EKF.

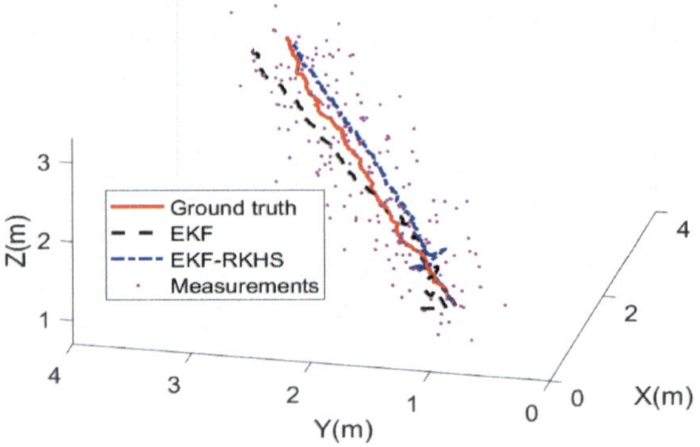

Fig. 5.1 Evolution of estimated trajectory in 3D space with EKF and EKF-RKHS for scenario I

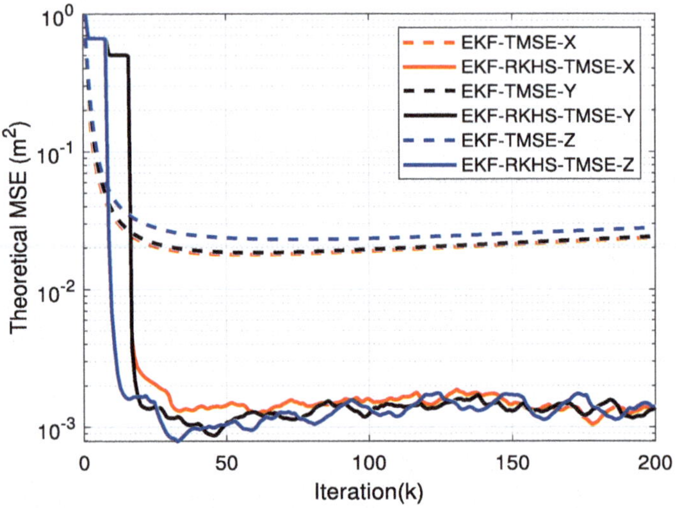

Fig. 5.2 Theoretical MSE along X-, Y-, and Z-axes for EKF and EKF-RKHS and corresponding to scenario I

Constant acceleration (scenario III): Specific to scenario III,

$$\mathbf{s}_0 = [1\text{m}, 2\text{m/s}, -2\text{m/s}^2, 1\text{m}, 1.2\text{m/s}, 2\text{m/s}^2, 1\text{m}, 1.7\text{m/s}, -1.5\text{m/s}^2]^T,$$

subsequently, $\mathbf{s}_k$ evolved using (5.3).

The comparative performance of EKF and EKF-RKHS is shown from Figs. 5.6 to 5.9 in terms of estimated trajectory and TMSEs. As depicted in Fig. 5.6, similar to scenarios I and II, the EKF-RKHS yields the estimated trajectory closer to the GT. Moreover, as shown in Fig. 5.7, unlike scenario I and scenario II, the TMSEs corresponding to the location of both EKF-RKHS and EKF are not converging.

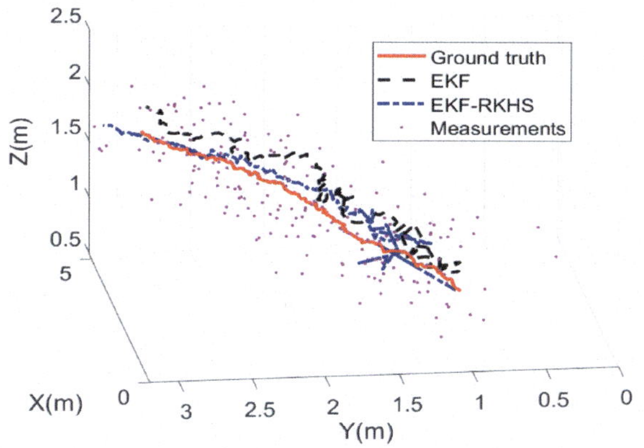

Fig. 5.3 Evolution of estimated trajectory in 3D space with EKF and EKF-RKHS for scenario II

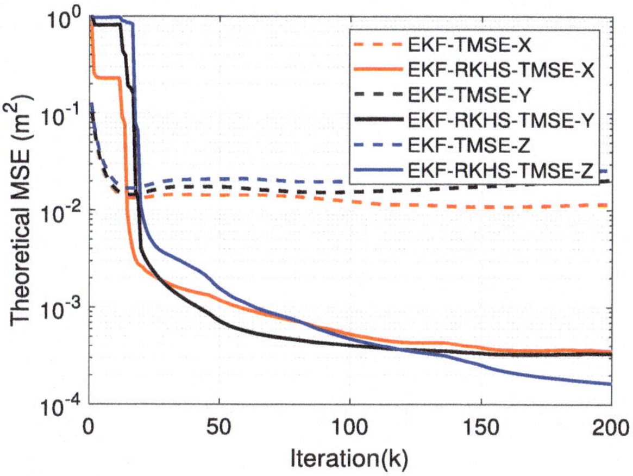

Fig. 5.4 Theoretical MSE along X-, Y-, and Z-axes for EKF and EKF-RKHS corresponding to scenario II

However, the TMSEs for EKF-RKHS along the x- and z-axes are lower than EKF. It is also observable that the TMSE for EKF-RKHS along the y-axis is converging at a higher value than EKF. The observed behavior of EKF-RKHS along the y-axis is due to positive acceleration, which causes fast movement along the y-axis.

Furthermore, Figs. 5.8 and 5.9 depict that the TMSEs for velocities and accelerations are also lower for EKF-RKHS and nearly converging compared to the TMSEs produced by EKF. In particular, as shown in Fig. 5.9, the TMSEs for accelerations along the x-, y-, and z-axes are not converging for EKF. In contrast, EKF-RKHS yields acceleration TMSEs not only lower than EKF but also nearly converging. In one way, this validates the suitability of using EKF-RKHS for the scenario where the random target motion hinders the accurate estimation of the state vector.

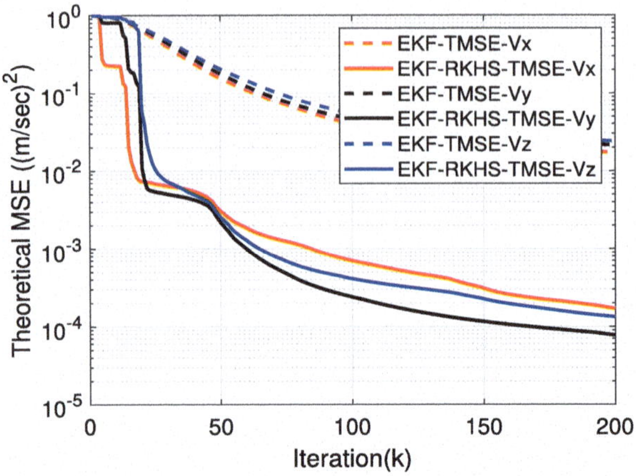

Fig. 5.5 Theoretical MSE for velocities in X-, Y-, and Z-axes for EKF and EKF-RKHS corresponding to scenario II

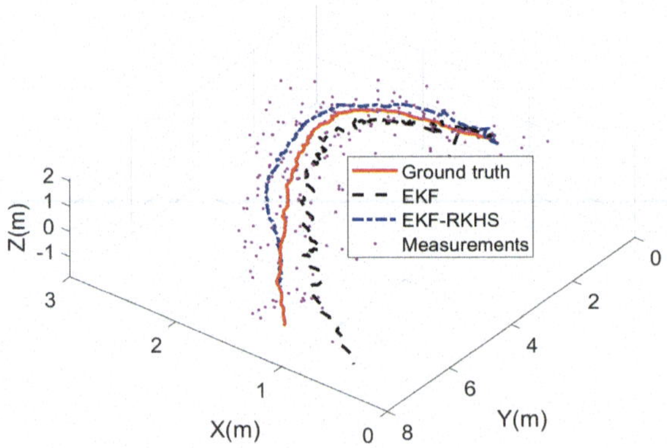

Fig. 5.6 Evolution of estimated trajectory in 3D space with EKF and EKF-RKHS for scenario III

In the preceding section, the EKF-RKHS is described in which **h** is estimated at each time instant with the KRLS algorithm [6], which is essentially an RKHS-based nonlinear estimation technique. Afterward, the estimated measurement function is plugged into EKF, making the EKF-RKHS independent of knowing the exact nonlinearity. In particular, the EKF-RKHS-based state estimator has been validated for different target motion models, and it has been shown that because of the estimation of the measurement function rather than approximation with Jacobian, EKF-RKHS achieves far better estimation accuracy than EKF. As mentioned, the theory of EKF and its advanced version, EKF-RKHS, is based on the assumption that the measurement follows the Gaussian distribution. Therefore, the EKF and hence the EKF-RKHS performance will drastically degrade subject to non-Gaussian distributed measurements. In [8], to make the EKF capable of dealing with the effects

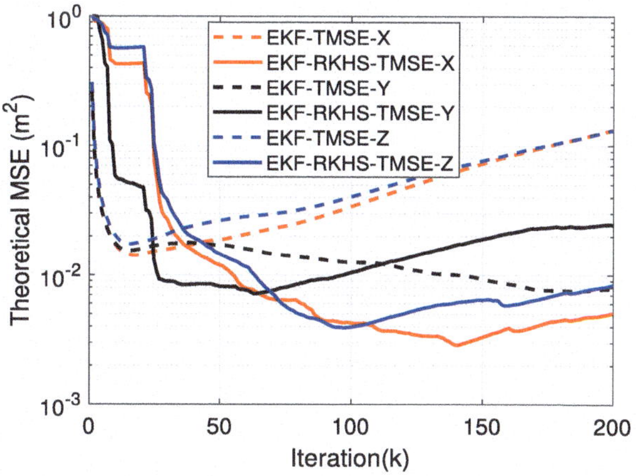

Fig. 5.7 Theoretical MSE for locations in X-, Y-, and Z-axes for EKF and EKF-RKHS corresponding to scenario III

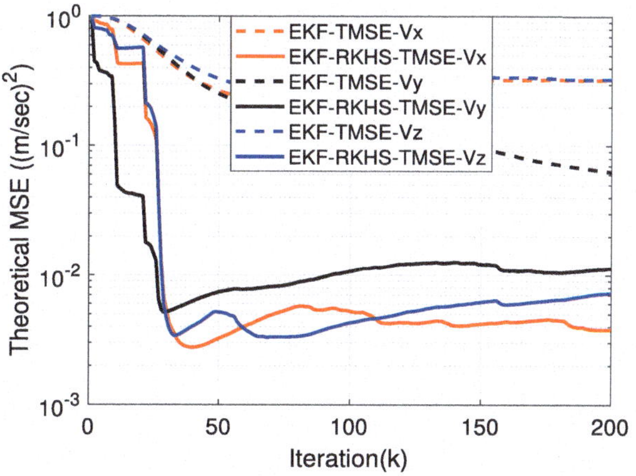

Fig. 5.8 Theoretical MSE for velocities in X-, Y-, and Z-axes for EKF and EKF-RKHS corresponding to scenario III

of non-Gaussianity, the EKF based on the MCC (EKF-MCC) has been introduced. The EKF utilizes the MMSE criterion to estimate unknown dynamic states and optimizes only second-order error statistics, essentially an error covariance. However, unlike EKF, EKF-MCC uses MCC, an information-theoretic criterion that considers the higher-order statistics for the estimation error. In EKF-MCC, the consideration of higher-order error statistics gives an upper hand over EKF to minimize the effect of non-Gaussianity.

The next section is devoted to the implementation of EKF-MCC-RKHS, which essentially makes EKF-MCC adaptive and eliminates the dependence of EKF-MCC

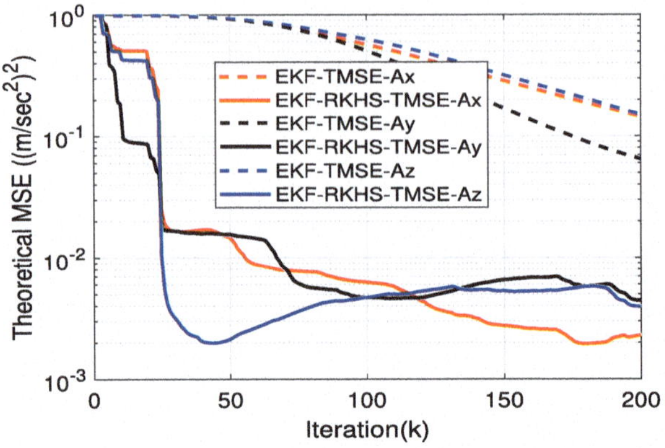

Fig. 5.9 Theoretical MSE for accelerations in X-, Y-, and Z-axes for EKF and EKF-RKHS corresponding to scenario III

on the knowledge of **h**. Similar to EKF-RKHS, EKF-MCC-RKHS first estimates **h** in RKHS using a well-known kernel recursive maximum correntropy (KRMC) algorithm, which is effectively a nonlinear adaptive algorithm in RKHS. Subsequently, the estimate of **h** replaces the Jacobian in EKF-MCC to perform further prediction and update.

5.4 Adaptive EKF Utilizing MCC

In this section, MCC is briefly discussed. Later, the EKF-MCC is introduced, followed by a detailed description of the EKF-MCC-RKHS.

5.4.1 Maximum Correntropy Criterion

The MCC maximizes the correntropy between two random variables $X \in \mathbb{R}$ and $Y \in \mathbb{R}$. If $\mathcal{P}_{X,Y}$ is the joint distribution function of X and Y, the correntropy between X and Y is given by

$$\mathcal{C}(X, Y) = \mathbb{E}(\kappa(X, Y)) = \int \kappa(x, y)\mathcal{P}_{X,Y}, \tag{5.14}$$

where $\kappa(\cdot, \cdot) \in \mathbb{R}$ is the Gaussian Mercer Kernel (GMK) and is defined by

$$\kappa(x, y) = \exp\left(- \frac{\|x - y\|^2}{2\sigma^2} \right),$$

where $\|\cdot\|$ is the l_2 norm and σ is the kernel width.

In practice, since $\mathcal{P}_{X,Y}$ is unknown, the $\mathbb{E}$ of the random variable is approximated by the limited samples available. Therefore, if N samples of X and Y are available, (5.14) can be estimated as

$$\hat{\mathcal{C}}(X, Y) = \frac{1}{N} \sum_{i=1}^{N} K_\sigma(e_i), \tag{5.15}$$

where $e_i = x_i - y_i$ is the error and $K_\sigma(e_i) = \kappa(e_i)$.

If Y is considered to be the estimate of X, then to produce the optimum value of Y closer to X, the MCC maximizes (5.15) as

$$Y = \hat{X} = \max_{y_i} \frac{1}{N} \sum_{i=1}^{N} K_\sigma(e_i). \tag{5.16}$$

In (5.16), $K_\sigma(e_i)$ can be expanded in Taylor's series as $K_\sigma(e_i) = 1 - \frac{e_i^2}{2\sigma^2} + \frac{e_i^4}{2\sigma^4} - \cdots$. Hence, it is explicit that (5.16) contains the higher-order terms of error or, more specifically, the MCC considers the higher-order statistics of the error in optimization. However, MMSE considers only the error-squared term (e_i^2) in the optimization. Considering higher-order statistics makes the MCC-based filter more robust against non-Gaussianity. In addition, σ in (5.16) prevents the MCC-based filter from exploding for high error. Based on this fact, next the EKF-MCC is introduced, which utilizes MCC to estimate $\mathbf{s}_k$ from non-Gaussian $\mathbf{y}_k$.

5.4.2 *EKF-MCC*

After prediction, mentioned in Sect. 5.3.1, the following equations give the Bayesian update equations of EKF-MCC:

$$\begin{aligned} \mathbf{y}_k^p &= \mathbf{H}_k \mathbf{s}_k^p, \\ \mathbf{K}_k^{MCC} &= \bar{\mathbf{P}}_k^p \mathbf{H}_k^T (\mathbf{H}_k \bar{\mathbf{P}}_k^p \mathbf{H}_k^T + \bar{\mathbf{R}}_w)^{-1}, \\ \hat{\mathbf{s}}_k &= \mathbf{s}_k^p + \mathbf{K}_k^{MCC} (\mathbf{y}_k - \mathbf{y}_k^p), \\ \mathbf{P}_k &= \bar{\mathbf{P}}_k^p - \mathbf{K}_k^{MCC} \mathbf{H}_k \bar{\mathbf{P}}_k^p, \end{aligned} \tag{5.17}$$

where $\mathbf{K}_k^{MCC}$ and $\bar{\mathbf{P}}_k^p$ are the Kalman gain and predicted error covariance matrix of EKF-MCC, respectively, and $\hat{\mathbf{s}}_k$ and $\mathbf{P}_k$ are the final estimate of $\mathbf{s}_k$ and updated error covariance matrix, respectively.

Furthermore, in the representation of $\mathbf{K}_k^{MCC}$, $\bar{\mathbf{P}}_k^p$ and $\bar{\mathbf{R}}_w$ are given by

$$\bar{\mathbf{P}}_k^p = \mathbf{C}_k^p \mathbf{E}_k^{s-1} \mathbf{C}_k^{pT},$$
$$\bar{\mathbf{R}}_w = \mathbf{C}_k^r \mathbf{E}_k^{y-1} \mathbf{C}_k^{rT},$$

where $\mathbf{C}_k^p$ and $\mathbf{C}_k^r$ are obtained from the Cholesky decomposition of $\mathbf{P}_k^p = \mathbf{C}_k^p \mathbf{C}_k^{pT}$ and $\mathbf{R}_w = \mathbf{C}_k^r \mathbf{C}_k^{rT}$, respectively, and

$$\mathbf{E}_k^s = diag[K_\sigma(e_1), K_\sigma(e_2), \ldots, K_\sigma(e_{n_s})],$$
$$\mathbf{E}_k^y = diag[K_\sigma(e_{n_s+1}), K_\sigma(e_{n_s+2}), \ldots, K_\sigma(e_{n_s+n_m})].$$

From (5.17), it can be inferred that in comparison to EKF's Kalman gain ($\mathbf{K}_k = \mathbf{P}_k^p \mathbf{H}_k^T (\mathbf{H}_k \mathbf{P}_k^p \mathbf{H}_k^T + \mathbf{R}_m)^{-1}$), the $\mathbf{K}_k^{MCC}$ provides a means to tackle the effect of non-Gaussianity with the help of σ. Also, for a high value of σ, $\mathbf{E}_k^s$ and $\mathbf{E}_k^y$ become the identity matrix, correspondingly, $\bar{\mathbf{P}}_k^p = \mathbf{P}_k^p$ and $\mathbf{K}_k^{MCC} = \mathbf{K}_k$, eventually EKF-MCC converges to EKF. In light of the above discussion, it is concluded that for a suitable value of σ, the EKF-MCC will have a different $\mathbf{K}_k^{MCC}$ to $\mathbf{K}_k$ of EKF, which makes the EKF-MCC suitable for dealing with non-Gaussianity.

However, in (5.17), it is explicitly stated that, like EKF, EKF-MCC also linearizes $\mathbf{h}(\cdot)$ via Jacobian, obtaining the first-order approximation of $\mathbf{h}(\cdot)$. Consequently, it restricts EKF-MCC from achieving high estimation accuracy and, in some cases, yielding inaccurate results when dealing with complex target motion models. Also, to obtain the Jacobian and, hence, to implement the next steps, EKF-MCC must know the exact form of $\mathbf{h}(\cdot)$ (which depends on the type of sensor and, in most cases, is not known a priori). The following subsection demonstrates an approach to deal with these shortcomings in EKF-MCC. The implementation of EKF-MCC-RKHS is described, which first estimates $\mathbf{h}(\cdot)$ in RKHS using a well-known KRMC algorithm, which is effectively a nonlinear adaptive algorithm in RKHS. Subsequently, the estimate of $\mathbf{h}(\cdot)$ replaces the Jacobian in EKF-MCC to perform further prediction and update.

5.4.3 EKF-MCC-RKHS

Following the basic mapping premises of EKF-RKHS, the estimate of $\mathbf{y}_k$ in $\mathcal{H}$ can be given by the well-known Representer's theorem [3] as

$$\hat{\mathbf{y}}_k = \langle \boldsymbol{\Omega}, \phi(\mathbf{s}_k) \rangle_{\mathcal{H}}. \tag{5.18}$$

In the search for the optimized value of $\boldsymbol{\Omega}$ and for the available k pairs of $\mathbf{y}_k$ and $\mathbf{s}_k$, $\{(\mathbf{s}_0, \mathbf{y}_0), \ldots, (\mathbf{s}_{k-1}, \mathbf{y}_{k-1})\}$, the KRMC algorithm maximizes the following cost function:

$$\mathcal{J}(\boldsymbol{\Omega}) = \max_{\boldsymbol{\Omega}} \sum_{j=0}^{k-1} \beta^{k-1-j} K_{\sigma_1}(\mathbf{y}_j - \hat{\mathbf{y}}_j) + \frac{1}{2}\beta^{k-1}\lambda\|\boldsymbol{\Omega}\|^2, \tag{5.19}$$

where β is the forgetting factor ($0 < \beta \leq 1$), introduced in (5.19) to enhance the effect of latest estimates, $K_{\sigma_1}(\cdot)$ is the GMK with kernel width σ_1, and λ is the regularization factor.

Utilizing (5.18), yields

$$\mathcal{J}(\boldsymbol{\Omega}) = \max_{\boldsymbol{\Omega}} \sum_{j=0}^{k-1} \beta^{k-1-j} K_{\sigma_1}(\mathbf{y}_j - \langle \boldsymbol{\Omega}, \phi(\mathbf{s}_j)\rangle_{\mathcal{H}}) + \frac{1}{2}\beta^{j}\lambda\|\boldsymbol{\Omega}\|^2. \tag{5.20}$$

Evaluating the gradient of (5.20) and equating it to zero the solution of $\boldsymbol{\Omega}$ is given by

$$\boldsymbol{\Omega} = \boldsymbol{\Phi}_{k-1}(\boldsymbol{\Phi}_{k-1}^T\boldsymbol{\Phi}_{k-1} + \lambda\beta^{k-1}\sigma_1^2\mathbf{B}_{k-1}^{-1})^{-1}\mathbf{y}_{k-1}, \tag{5.21}$$

where $\boldsymbol{\Phi}_{k-1} = [\phi(\mathbf{s}_0), \ldots, \phi(\mathbf{s}_{k-1})]^T$, and

$$\begin{aligned}\mathbf{B}_{k-1} = diag[&\beta^{k-1}K_{\sigma_1}(\mathbf{y}_0 - \langle \boldsymbol{\Omega}, \phi(\mathbf{s}_0)\rangle_{\mathcal{H}}), \ldots,\\ &K_{\sigma_1}(\mathbf{y}_{k-1} - \langle \boldsymbol{\Omega}, \phi(\mathbf{s}_{k-1})\rangle_{\mathcal{H}})].\end{aligned}$$

Substituting (5.21) in (5.18) yields

$$\hat{\mathbf{y}}_k = [\langle \phi(\mathbf{s}_0), \phi(\mathbf{s}_k)\rangle_{\mathcal{H}}, \ldots, \langle \phi(\mathbf{s}_{k-1}), \phi(\mathbf{s}_k)\rangle_{\mathcal{H}}]\mathbf{a}_{k-1}^T, \tag{5.22}$$

where $\mathbf{a}_{k-1} = (\boldsymbol{\Phi}_{k-1}^T\boldsymbol{\Phi}_{k-1} + \lambda\beta^{k-1}\sigma_1^2\mathbf{B}_{k-1}^{-1})^{-1}$.

With an analogy of (5.22) with (5.2), it can be infer that the estimate of $\mathbf{h}$ at kth time instant is given by

$$\hat{\mathbf{h}}_k = [\langle \phi(\mathbf{s}_0), \phi(\mathbf{s}_k)\rangle_{\mathcal{H}}, \ldots, \langle \phi(\mathbf{s}_{k-1}), \phi(\mathbf{s}_k)\rangle_{\mathcal{H}}]. \tag{5.23}$$

In (5.23), $\langle \phi(\mathbf{s}_i), \phi(\mathbf{s}_j)\rangle_{\mathcal{H}}$ represents the inner product of $\phi(\mathbf{s}_i)$ and $\phi(\mathbf{s}_j)$ in $\mathcal{H}$. Since $\phi(\cdot)$ is inaccessible, (5.12) can be simplified by evoking the celebrated Mercer theorem [7]. Mercer's theorem states that the inner product in $\mathcal{H}$ is efficiently calculated in Euclidean space using $\kappa(\cdot, \cdot)$.

Hence, utilizing Mercer's theorem, the simplified form of $\hat{\mathbf{h}}_k$ is given by

$$\hat{\mathbf{h}}_k = [\kappa(\mathbf{s}_0, \mathbf{s}_k), \ldots, \kappa(\mathbf{s}_{k-1}, \mathbf{s}_k)]. \tag{5.24}$$

The (5.24) suggests that, unlike EKF-MCC, the knowledge of $\mathbf{h}$ is no longer needed. Instead, $\mathbf{h}$ can be estimated iteratively along with the prediction and update step of EKF-MCC. Moreover, at the instant kth, $\hat{\mathbf{h}}_k$ is based on the current state vector $\mathbf{s}_k$ and the past state vectors $\{\mathbf{s}_0, \ldots, \mathbf{s}_{k-1}\}$. This implies that at kth instant after obtaining $\hat{\mathbf{s}}_k$ from EKF-MCC, in (5.24), $\hat{\mathbf{s}}_k$ replaces $\mathbf{s}_k$. Afterward, $\hat{\mathbf{h}}_k$ replaces $\mathbf{h}$ in EKF-MCC to perform the further prediction and update steps. The process repeats iteratively for $k = 1, 2, \ldots, K$. In addition, σ_1 in $\mathbf{a}_{k-1}$ provides additional freedom to adjust the performance of EKF-MCC-RKHS against non-Gaussianity.

The advantages of using EKF-MCC-RKHS over EKF-MCC and EKF are threefold: (1) Unlike EKF-MCC and EKF, the implementation of EKF-MCC-RKHS is not limited to knowing the exact form of $\mathbf{h}$. (2) Since, unlike EKF-MCC and EKF, the estimate of $\mathbf{h}$ is used, EKF-MCC-RKHS will produce the estimate of $\mathbf{s}_k$ with higher precision. (3) Finally, to estimate $\mathbf{h}$, since MCC is used in KRMC, like EKF-MCC, EKF-MCC-RKHS is suitable to use with non-Gaussianity; however, it provides a better estimation accuracy than EKF-MCC.

The pseudo-algorithm for EKF-MCC-RKHS is given in Algorithm 2, where $\mathbf{z}_k$, r_k, $\mathbf{Q}_k$, $\boldsymbol{\theta}_k$, $\mathbf{e}_k$, and $\mathbf{a}_k$ are as per [17].

Algorithm 2: Implementation of EKF-MCC-RKHS

Intialization :

$\mathbf{s}_0, \mathbf{y}(\mathbf{s}_0), \hat{\mathbf{s}}_0, \mathbf{P}_0, \hat{\mathbf{h}}_0 = [\kappa(\hat{\mathbf{s}}_0, \hat{\mathbf{s}}_0)], \mathbf{Q}_0 = (\lambda\beta\sigma_1^2 + \kappa(\hat{\mathbf{s}}_0, \hat{\mathbf{s}}_0))^{-1}, \mathbf{a}_0 = [1, \ldots, 1]^{1\times n_m}$

for $k = 1, 2, 3, \cdots, K$ **do**

EKF-MCC

Compute $\mathbf{s}_k^p$ and $\mathbf{P}_k^p$ using (5.28)

$\mathbf{y}_k^p = \mathbf{a}_{k-1}^T \hat{\mathbf{h}}_{k-1}^T$

$$\hat{\mathbf{H}}_k = \left\{ \left(\frac{\partial \hat{\mathbf{h}}_{k-1}}{\partial \hat{\mathbf{s}}_{k-1}} \bigg|_{\hat{\mathbf{s}}_{k-1} = \mathbf{s}_k^p} \right) \mathbf{a}_{k-1} \right\}^T$$

Compute $\mathbf{K}_k^{MCC}$, $\hat{\mathbf{s}}_k$, and $\mathbf{P}_k$ using (5.29)

KRMC

$\hat{\mathbf{h}}_k = [\kappa(\hat{\mathbf{s}}_0, \hat{\mathbf{s}}_k), \ldots, \kappa(\hat{\mathbf{s}}_{k-1}, \hat{\mathbf{s}}_k)]$

$\mathbf{e}_k = \mathbf{y}_k - \mathbf{a}_{k-1}^T \hat{\mathbf{h}}_{k-1}^T$

$\mathbf{z}_k = \mathbf{Q}_{k-1} \hat{\mathbf{h}}_{k-1}^T$

$\theta_k = (\exp(-\frac{\|\mathbf{e}_k\|^2}{2\sigma_1^2}))^{-1}$

$r_k = \lambda\beta^k\sigma_1^2\theta_k + \kappa(\hat{\mathbf{s}}_k, \hat{\mathbf{s}}_k) - \mathbf{z}_k^T \hat{\mathbf{h}}_{k-1}^T$

$$\mathbf{Q}_k = r_k^{-1} \begin{bmatrix} \mathbf{Q}_{k-1} r_k + \mathbf{z}_k \mathbf{z}_k^T & -\mathbf{z}_k \\ -\mathbf{z}_k^T & 1 \end{bmatrix}$$

$$\mathbf{a}_k = \begin{bmatrix} \mathbf{a}_{k-1} - r_k^{-1} \mathbf{z}_k \mathbf{e}_k^T \\ r_k^{-1} \mathbf{e}_k^T \end{bmatrix}$$

end

5.4.4 *Simulation Results and Analysis*

This section provides a detailed description of the simulation results performed to evaluate the performance of the EKF-MCC-RKHS compared to EKF-MCC and EKF. The simulations are characterized in Scenario I and Scenario II. In Scenario I, the simulations are performed to validate the better performance of EKF-MCC over EKF. Since EKF and EKF-RKHS use the same MMSE criterion, in Section I, the EKF-MCC is compared with EKF only. Henceforth, Scenario II evaluates the performance of EKF-MCC-RKHS compared to its counterparts EKF-MCC and EKF. For both scenarios, $\mathbf{y}_k$ is assumed to be in spherical coordinates and follow the Gaussian mixture with $\mathbf{w}_k \in p_1\mathcal{N}_{\mathbb{R}}(\mathbf{0}, \mathbf{R}_w^1) + p_2\mathcal{N}_{\mathbb{R}}(\mathbf{0}, \mathbf{R}_w^2)$, correspondingly, $\mathbf{R}_w = \sum_{g=1}^{2} p_g \mathbf{R}_w^g$. The $p1 = 0.1$ and $p2 = 0.9$, the $\mathbf{R}_w^1 = \sigma_{w1}^2 \mathbf{I}_{n_m}$ and $\mathbf{R}_w^2 = \sigma_{w2}^2 \mathbf{I}_{n_m}$, where $\sigma_{w1}^2 = 0.15$ and $\sigma_{w2}^2 = 20$, respectively. Scenario I and Scenario II are both simulated for $K = 150$ CPIs, i.e., $k = 1, 2, \ldots, 150$, $T_{cpi} = 0.01$ s, and $\mathbf{Q}_u$ according to (5.37), where $\sigma_a = 10^{-2}$. The $\hat{\mathbf{s}}_0$ and $\mathbf{P}_0$ for both scenarios are considered as $[1, 1, \ldots, 1]^{n_s \times 1}$ and $\mathbf{I}_{n_s}$, respectively. Furthermore, since $\mathbf{y}_k$ is available in spherical coordinate ($\mathbf{y}_k = [\hat{r}_k, \hat{\theta}_k, \hat{\phi}_k]^T$), the associated nonlinear $\mathbf{h}(\cdot)$ is given in Sect. 5.3.3.

The target is assumed to move in the 3D space in an indoor scene. Consequently, the maximum distance traveled by the target in the x-, y-, and z-directions is limited to 10 m. The performance of the filters for Scenarios I and II is quantified with the root mean square error (RMSE) along the positions, velocities, and accelerations in the x-, y-, and z-directions. The RMSEs for the elements of $\hat{\mathbf{s}}_k$ are defined as

$$\text{RMSE}(i) = \sqrt{\frac{1}{K}\sum_{k=0}^{K-1}(\mathbf{s}_k(i) - \hat{\mathbf{s}}_k(i))^2}; \quad i = 1, 2, \ldots, n_s, \tag{5.25}$$

where i denotes the ith element of $\mathbf{s}_k$ and $\hat{\mathbf{s}}_k$.

Scenario I: In Scenario I, simulations are performed to evaluate the performance of EKF and EKF-MCC. For EKF-MCC, the $\sigma = 0.95$. The GT and the estimated trajectories obtained from EKF and EKF-MCC are shown in Fig. 5.10. As depicted in Fig. 5.10, the trajectory estimated by EKF-MCC is close to GT. The improved performance of EKF-MCC over EKF in the presence of MCC is shown in Fig. 5.11. In Fig. 5.11, the RMSE obtained by EKF-MCC and EKF to estimate all parameters $n_s = 9$ is shown. Referring to Figs. 5.10 and 5.11, it was concluded that the EKF-MCC outperforms EKF when the system is affected by non-Gaussian noise.

Scenario II: Simulations to evaluate the comparative performance of EKF-MCC and EKF-MCC-RKHS in Scenario II. It is already shown in Figs. 5.14 and 5.16 that EKF-MCC outperforms EKF. However, for better clarity, in this subsection, the performance of EKF-MCC-RKHS is compared with that of EKF-MCC and EKF both. In the simulations, for EKF-MCC $\sigma = 0.95$ and for KRMC $\sigma_1 = 5$, also β and λ are 1 and 0.004, respectively. Furthermore, it is depicted in Fig. 5.12 that EKF-MCC-RKHS produces the estimated trajectory in close proximity to GT. Also, from Fig. 5.12, it can be inferred that EKF-MCC performs better than EKF with system non-Gaussianity; however, because of estimating $\mathbf{h}(\cdot)$ and using MCC, EKF-MCC-RKHS performs better than both EKF-MCC and EKF. Furthermore, Fig. 5.13 validates the improved performance of EKF-MCC-RKHS over EKF-MCC and EKF in terms of RMSE of all estimated parameters.

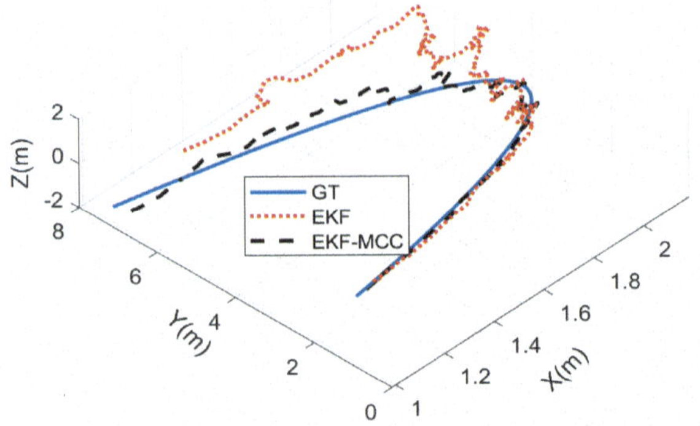

Fig. 5.10 Evolution of GT and estimated trajectories in 3D space with EKF and EKF-MCC for Scenario I

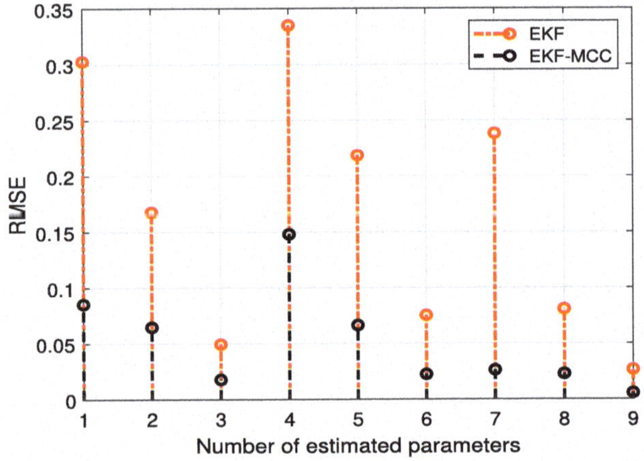

Fig. 5.11 RMSE in the estimation of all parameters (x, v_x, a_x, y, v_y, a_z, z, v_z, and a_z) with EKF and EKF-MCC for Scenario I

In the previous section, an adaptive version of the MCC-based EKF, known as EKF-MCC-RKHS, was introduced. It was observed that by incorporating MCC, EKF-MCC-RKHS produces a more accurate estimate of **h**, which in turn improves the estimation of target kinematic parameters. In addition to MCC, another non-MMSE criterion is the MEE criterion. MEE is a powerful learning criterion that has proven effective in applications such as robust regression, classification, system identification, and adaptive filtering. Numerous experimental results indicate that MEE can outperform MCC in various scenarios, although it comes with slightly higher computational complexity. As a result, in the next section, following the approach used for EKF-RKHS and EKF-MCC-RKHS, the adaptive version of the MEE-based EKF, called EKF-KRMEE, is introduced. Unlike EKF-RKHS and EKF-

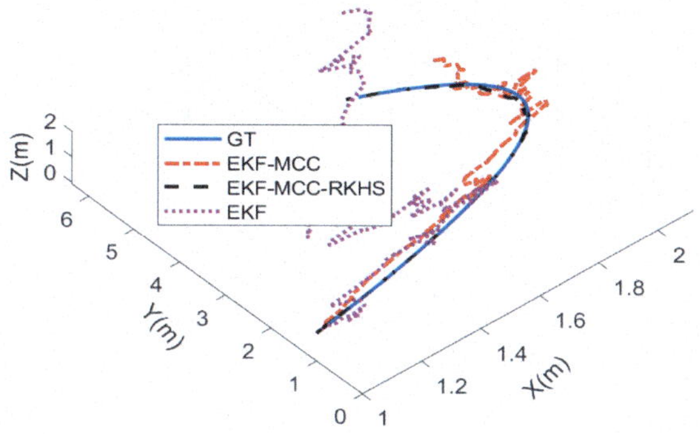

Fig. 5.12 Evolution of GT and estimated trajectories in 3D space with EKF, EKF-MCC, and EKF-MCC-RKHS for Scenario II

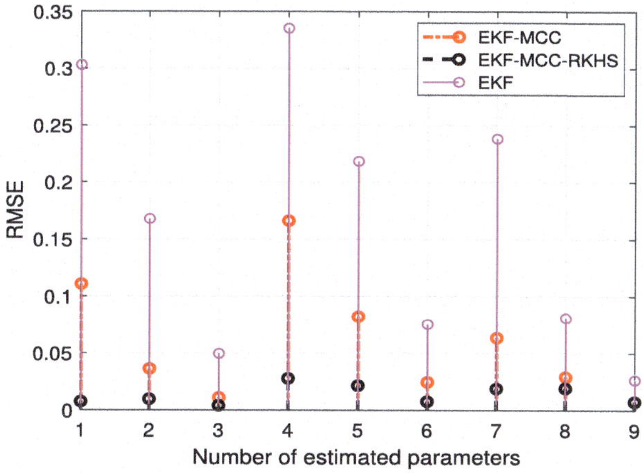

Fig. 5.13 RMSE in the estimation of all parameters (x, v_x, a_x, y, v_y, a_z, z, v_z, and a_z) with EKF, EKF-MCC, and EKF-MCC-RKHS for Scenario II

MCC-RKHS, EKF-KRMEE utilizes MEE to address the challenges posed by non-Gaussian noise.

5.5 Adaptive EKF Utilizing MEE

In this section, the MEE criterion is described in detail. Subsequently, the implementation of an MEE-based EKF is discussed. Estimation of $\mathbf{h}(\cdot)$, using the adaptive version of the MEE-based EKF: EKF-KRMEE is discussed next.

5.5.1 Minimum Error Entropy Criterion

Unlike MMSE and MCC, MEE minimizes the information contained in the error ($e \in \mathbb{R}$). The information contained in e is measured by Renyi's entropy defined as

$$H_\alpha(e) = \frac{1}{(1-\alpha)} \log \mathbb{E}[p^{\alpha-1}(e)], \tag{5.26}$$

where α is the entropy order, e is the error in the prediction of some random variable $Y \in \mathbb{R}$ as $\hat{f}(X) \in \mathbb{R}$, i.e., $e = Y - \hat{f}(X)$, and $p(e)$ is the distribution function of e.

Instead of minimizing (5.26), the specialized form of Renyi's entropy defined in (5.26) is considered for $\alpha = 2$, this results in the famous Shannon's entropy, defined as

$$H_2(e) = -\log \mathbb{E}[p(e)]. \tag{5.27}$$

Subsequently, if the L samples of Y and X are available as $(x_i, y_i)\forall i = 1, \dots, L$, then $e_i = (y_i - \hat{f}(x_i))\forall i = 1, \dots, L$ and $p(e)$ are estimated using the kernel $G_\sigma(e - e_i)$ as

$$\hat{p}(e) = \frac{1}{L}\sum_{i=1}^{L} G_\sigma(e - e_i), \tag{5.28}$$

where $G_\sigma(e - e_i)$ is the most widely used Gaussian kernel function with width σ, i.e., $G_\sigma(e - e_i) = \frac{1}{\sqrt{(2\pi)}\sigma} \exp(-\frac{(e-e_i)^2}{2\sigma^2})$.

Referring to (5.27), it is evident that minimizing $H_2(e)$ is equivalent to maximizing the argument ($\mathbb{E}[p(e)]$). Consequently, MEE maximizes $\mathbb{E}[p(e)]$, which can be approximated as

$$\mathcal{J}(X) = \frac{1}{L}\sum_{j=1}^{L} p(e_j). \tag{5.29}$$

Substituting (5.28) into (5.29) yields

$$\mathcal{J}(X) = \frac{1}{L^2}\sum_{j=1}^{L}\sum_{i=1}^{L} G_\sigma(e_j - e_i). \tag{5.30}$$

The MEE maximizes (5.29) to yield $\hat{f}(X)$ closer to Y as

$$Y \approx \hat{f}(X) = \max_{f(\cdot)\in\mathbb{R}} \frac{1}{L^2}\sum_{j=1}^{L}\sum_{i=1}^{L} G_\sigma(e_j - e_i). \tag{5.31}$$

In Eq. (5.31), the term $G_\sigma(e_j - e_i)$ can be expanded using Taylor's series as $G_\sigma(e_j - e_i) = 1 - \frac{(e_j-e_i)^2}{2\sigma^2} + \frac{(e_j-e_i)^4}{2\sigma^4} - \cdots$. Consequently, Eq. (5.31) incorporates higher-order terms of error differences. Specifically, unlike MCC, MEE takes into account higher-order error difference statistics in its estimation process. This consideration of higher-order statistics enhances the robustness of the MEE-based filter against non-Gaussianity compared to MCC and MMSE. Furthermore, MEE employs the error difference $(e_j - e_i)$, providing MEE with an advantage over MCC, which only considers e_i. Beyond these crucial distinctions from MMSE and MCC, the parameter σ prevents the MEE-based filter from becoming unstable for high errors, a situation that commonly arises with multimodal distributed noises. With these points established, the next step involves describing the MEE-based EKF, which uses MEE to estimate $\mathbf{s}_k$ from non-Gaussian $\mathbf{y}_k$.

5.5.2 MEE-Based EKF

After the prediction, mentioned in Sect. 5.2, in the update step, the MEE-based EKF invokes the MEE criterion, and the predictions made in (5.5) are updated with the help of available measurements from the sensor ($\mathbf{y}_k$). The following equations give the updated equations of MEE-based EKF:

$$\begin{aligned} \mathbf{y}_k^p &= \mathbf{H}_k \mathbf{s}_k^p, \\ \mathbf{K}_k^{MEE} &= \left[\bar{\mathbf{P}}_{k-1}^p + \mathbf{H}_k^T \bar{\mathbf{SY}}_{k-1} + (\bar{\mathbf{YS}}_{k-1} + \mathbf{H}^T \bar{\mathbf{R}}_m \mathbf{H})\right]^{-1} \\ &\quad \times ((\bar{\mathbf{YS}}_{k-1} + \mathbf{H}^T \bar{\mathbf{R}}_m)), \\ \hat{\mathbf{s}}_k &= \mathbf{s}_k^p + \mathbf{K}_k^{MEE}(\mathbf{y}_k - \mathbf{y}_k^p), \\ \mathbf{P}_k &= \bar{\mathbf{P}}_k^p - \mathbf{K}_k^{MEE} \mathbf{H}_k \bar{\mathbf{P}}_k^p, \end{aligned} \tag{5.32}$$

where $\mathbf{K}_k^{MEE}$, $\bar{\mathbf{P}}_k^p$, $\bar{\mathbf{SY}}_{k-1}$, and $\bar{\mathbf{YS}}_{k-1}$ are the Kalman gain, the predicted error covariance matrix, the cross-covariance matrix of $\mathbf{s}_k$ and $\mathbf{y}_k$, the cross-covariance matrix of $\mathbf{y}_k$ and $\mathbf{s}_k$ of the MEE-based EKF, respectively, and $\hat{\mathbf{s}}_k$ and $\mathbf{P}_k$ are the final estimate of $\mathbf{s}_k$ and the updated error covariance matrix, respectively.

From (5.32), it can be inferred that in comparison to EKF's Kalman gain ($\mathbf{K}_k = \mathbf{P}_k^p \mathbf{H}_k^T (\mathbf{H}_k \mathbf{P}_k^p \mathbf{H}_k^T + \mathbf{R}_m)^{-1}$), the $\mathbf{K}_k^{MEE}$ provides a means to tackle the effect of non-Gaussianity with the help of σ. In fact, for a suitable value of σ, $\mathbf{K}_k^{MEE}$ differs from $\mathbf{K}_k$ and provides robustness against non-Gaussianity. However, similar to the EKF and the MCC-based EKF, the MEE-based EKF also makes a linear approximation of $\mathbf{h}$ via Jacobian. This approximation causes the MEE-based EKF to perform worse, especially when dealing with complex measurement models. Therefore, this subsection presents a unified approach of clubbing MEE-based EKF with the KRMEE algorithm and introduces a new EKF-KRMEE filter. The EKF-KRMEE first estimates $\mathbf{h}$ in RKHS using the KRMEE algorithm, subsequently, $\hat{\mathbf{h}}$ replaces the Jacobian in MEE-based EKF for further prediction and update.

5.5.3 EKF-KRMEE

In this subsection, the use of the KRMEE algorithm to obtain $\hat{\mathbf{h}}_k$ is demonstrated. Similar to EKF-RKHS and EKF-MCC-RKHS after the basic mapping of $\mathbf{y}_k$, the estimate of $\mathbf{y}_k$ in $\mathcal{H}$ can be given by the inner product ($\langle\cdot,\cdot\rangle_{\mathcal{H}}$) of $\mathbf{y}_k$ and the unknown weight matrix $\boldsymbol{\Omega}$ as

$$\hat{\mathbf{y}}_k = \langle \boldsymbol{\Omega}^T, \phi(\mathbf{s}_k)\rangle_{\mathcal{H}}. \tag{5.33}$$

To reduce the estimation error ($\mathbf{e}_k = \mathbf{y}_k - \hat{\mathbf{y}}_k$) s.t. an optimized value of $\boldsymbol{\Omega}$ is obtained, the KRMEE algorithm maximizes the cost function defined in (5.31) as

$$\mathcal{J}(\boldsymbol{\Omega}) = \max_{\boldsymbol{\Omega}\in\mathcal{H}} \frac{1}{L^2}\sum_{j=1}^{L}\sum_{i=1}^{L}\gamma^{i+j}G_\sigma(e_j - e_i) - \frac{1}{2}\lambda\|\boldsymbol{\Omega}\|^2, \tag{5.34}$$

where γ is the forgetting factor s.t ($0 < \gamma \leq 1$), introduced in (5.34) to enhance the effect of latest estimates, and λ is the regularization factor.

Equating the gradient of (5.34) w.r.t Ω and equating it to zero yields the following solution:

$$\Omega = \boldsymbol{\Phi}_L(\boldsymbol{\Phi}_L^T\boldsymbol{\Phi}_L + \sigma^2\gamma\mathbf{B}_L^{-1})^{-1}\mathbf{Y}_L, \tag{5.35}$$

where

$$\begin{cases} \gamma = \frac{L^2\lambda}{2}, \\ \Phi_L = [\phi(s_1), \phi(s_2), \ldots, \phi(s_L)], \\ \mathbf{Y}_L = [\mathbf{y}_1, \mathbf{y}_2, \ldots, \mathbf{y}_L]^T, \\ \mathbf{B}_L = \mathbf{M}_L - \mathbf{N}_L, \\ \mathbf{M}_L(i,j) = \gamma^{i+j}G_\sigma(e_j - e_i), \forall i, j = 1, \ldots, L, \\ \mathbf{N}_L(i,j) = \begin{cases} \sum_{k=1}^{L}\gamma^{i+k}G_\sigma(e_j - e_k), i = j \\ \mathbf{0}, i \neq j \end{cases} \end{cases}$$

The (5.35) can be represented as $\Omega = \boldsymbol{\Phi}_L\mathbf{a}_L$, where $\mathbf{a}_L = (\boldsymbol{\Phi}_L^T\boldsymbol{\Phi}_L + \sigma^2\gamma\mathbf{B}_L^{-1})^{-1}$ $\mathbf{Y}_L$, subsequently, substituting (5.35) into (5.33) yields

$$\hat{\mathbf{y}}_k = [\langle\phi(\mathbf{s}_1^T), \phi(\mathbf{s}_k)\rangle_{\mathcal{H}}, \ldots, \langle\phi(\mathbf{s}_L^T), \phi(\mathbf{s}_k)\rangle_{\mathcal{H}}]\mathbf{a}_L^T. \tag{5.36}$$

Comparing (5.36) with (5.1), infer that $\hat{\mathbf{h}}(\cdot)$ is given by

$$\hat{\mathbf{h}}(\mathbf{s}_k) = [\langle\phi(\mathbf{s}_1^T), \phi(\mathbf{s}_k)\rangle_{\mathcal{H}}, \ldots, \langle\phi(\mathbf{s}_L^T), \phi(\mathbf{s}_k)\rangle_{\mathcal{H}}]. \tag{5.37}$$

In (5.37), the computation of $\langle\phi(\mathbf{s}_L^T), \phi(\mathbf{s}_k)\rangle_{\mathcal{H}}$ requires the knowledge of $\phi(\cdot)$. However, $\phi(\cdot)$ is an implicit mapping in $\mathcal{H}$ whose exact form is unknown. Therefore, to compute (5.37), Mercer's theorem is evoked, which states that

$$\langle\phi(\mathbf{s}_L^T), \phi\mathbf{s}_k)\rangle_{\mathcal{H}}$$

could be effectively computed in the Euclidean space via Gaussian kernel $\kappa(\cdot,\cdot)$ as

$$\hat{\mathbf{h}}(\mathbf{s}_k) = [\kappa(\mathbf{s}_1, \mathbf{s}_k), \ldots, \kappa(\mathbf{s}_l, \mathbf{s}_k), \ldots, \kappa(\mathbf{s}_L, \mathbf{s}_k)], \tag{5.38}$$

where

$$\langle \phi(\mathbf{s}_l^T), \phi(\mathbf{s}_k) \rangle_{\mathcal{H}} = \kappa(\mathbf{s}_l, \mathbf{s}_k) = \frac{1}{\sqrt{(2\pi)}\sigma_1} \exp\left(- \frac{\|\mathbf{s}_l - \mathbf{s}_k\|^2}{2\sigma_1^2} \right).$$

From (5.38), it can be easily inferred that, contrary to MEE-based EKF, the exact form of $\mathbf{h}$ is no longer required. Instead, $\mathbf{h}$ can be estimated iteratively using a KRMEE-based adaptive algorithm. Furthermore, $\mathbf{h}$ can be calculated in conjunction with the prediction and update steps of MEE-based EKF. Therefore, the combination of MEE-based EKF and the KRMEE adaptive algorithm results in the EKF-KRMEE adaptive filter.

The EKF-KRMEE is beneficial over MEE-based EKF and EKF for the following reasons: (1) the implementation of EKF-KRMEE is no longer restricted to knowing the exact form of $\mathbf{h}$, (2) using the estimate of $\mathbf{h}$ in place of the Jacobian makes EKF-KRMEE a suitable choice to estimate $\mathbf{s}_k$, and (3) lastly, since the MEE criterion is evoked, EKF-KRMEE is suitable to use in non-Gaussian scenarios and consequently provides better estimation accuracy than MEE-based EKF and MCC-based EKF.

The pseudo-algorithm for EKF-KRMEE is given in Algorithm 3, where $\mathbf{z}_k$, r_k, $\mathbf{Q}_k$, $\boldsymbol{\theta}_k$, $\mathbf{e}_k$, and $\mathbf{a}_k$ are as per [16].

5.5.4 Simulation Results and Inferences

In this section, the performance of EKF-KRMEE is validated through computer simulations. Firstly, the performance of MEE-based EKF is compared with its counterparts, namely, MCC-based EKF and EKF, under non-Gaussian measurement. Subsequently, on the basis of the results, the conclusion regarding the superiority of MEE-based EKF over MCC-based EKF and EKF is drawn. Later, the adaptive EKF-KRMEE filter is compared with its non-adaptive counterpart, the MEE-based EKF. Given that MEE-based EKF outperforms EKF- and MCC-based EKF, it implies that their adaptive versions, EKF-RKHS and EKF-MCC-RKHS, respectively, will also be surpassed by MEE-based EKF. It is essential to note that the primary goal of this work is to introduce an adaptive algorithm for MEE-based EKF, aiming to enhance the state estimation accuracy of MEE-based EKF in non-Gaussian scenarios. Consequently, in the second part of the simulation, the adaptive EKF-KRMEE filter is compared solely with the MEE-based EKF.

Furthermore, to account for non-Gaussianity in $\mathbf{y}_k$, $\mathbf{w}_k$ is modeled in the manner previously described in Sect. 5.4.4. Out of numerous non-Gaussianity, the Gaussian mixture is chosen to check the performance of MEE-based filter for multimodal distribution. The simulations are performed for $K = 150$ CPIs, i.e., $k = 1, 2, \ldots, 150$,

Algorithm 3: Implementation of EKF-KRMEE

Intialization :

$\mathbf{s}_0, \mathbf{y}(\mathbf{s}_0), \hat{\mathbf{s}}_0, \mathbf{P}_0, \hat{\mathbf{h}}_0 = [\kappa(\hat{\mathbf{s}}_0, \hat{\mathbf{s}}_0)], \mathbf{Q}_0 = (\lambda\sigma^2 + \kappa(\hat{\mathbf{s}}_0, \hat{\mathbf{s}}_0))^{-1}, \mathbf{a}_0 = \mathbf{Q}_0\mathbf{y}_0^T$

for $k = 1, 2, 3, \cdots, K$ **do**

EKF-MEE

Compute $\mathbf{s}_k^p$ and $\mathbf{P}_k^p$ using (5.28)

$\mathbf{y}_k^p = \mathbf{a}_{k-1}^T \hat{\mathbf{h}}_{k-1}^T$

$\hat{\mathbf{H}}_k = \left\{ \left(\frac{\partial \hat{\mathbf{h}}_{k-1}}{\partial \hat{\mathbf{s}}_{k-1}} \Big|_{\hat{\mathbf{s}}_{k-1}=\mathbf{s}_k^p} \right) \mathbf{a}_{k-1} \right\}^T$

Compute $\mathbf{K}_k^{MEE}$, $\hat{\mathbf{s}}_k$, and $\mathbf{P}_k$ using (5.29)

KRMEE

$\hat{\mathbf{h}}_k = [\kappa(\hat{\mathbf{s}}_0, \hat{\mathbf{s}}_k), \ldots, \kappa(\hat{\mathbf{s}}_L, \hat{\mathbf{s}}_k)]$

$\mathbf{e}_k = \mathbf{y}_k - \mathbf{a}_{k-1}^T \hat{\mathbf{h}}_k^T$

$\mathbf{z}_k = \mathbf{Q}_{k-1} \hat{\mathbf{h}}_{k-1}^T$

$\phi_k = \sum_{i=1}^{L} \gamma^{i+L} G_\sigma(e_k - e_i)$,

$r_k = \gamma_2 \sigma^2 \phi_k + \kappa(\hat{\mathbf{s}}_k, \hat{\mathbf{s}}_k) - \mathbf{z}_k^T \hat{\mathbf{h}}_{k-1}^T$

$\mathbf{Q}_k = r_k^{-1} \begin{bmatrix} \mathbf{Q}_{k-1} r_k + \mathbf{z}_k \mathbf{z}_k^T & -\mathbf{z}_k \\ -\mathbf{z}_k^T & 1 \end{bmatrix}$

$\mathbf{a}_k = \begin{bmatrix} \mathbf{a}_{k-1} - r_k^{-1} \mathbf{z}_k \mathbf{e}_k^T \\ r_k^{-1} \mathbf{e}_k^T \end{bmatrix}$

end

$T_{cpi} = 0.01$ s, and $\mathbf{Q}_u$ is according to (5.26), where $\sigma_a = 10^{-2}$. The $\hat{\mathbf{s}}_0$ and $\mathbf{P}_0$ for both scenarios are considered as $[1, 1, \ldots, 1]^{n_s \times 1}$ and $\mathbf{I}_{n_s}$, respectively. The $\mathbf{y}_k$ is assumed to be available in spherical coordinate ($\mathbf{y}_k = [\hat{r}_k, \hat{\theta}_k, \hat{\phi}_k]^T$). Consequently, the associated true form of the nonlinear function: $\mathbf{h}(\cdot)$ is the same as that mentioned in Sect. 5.3.3.

The target is assumed to move in an indoor setting. Consequently, the maximum distance traveled by the target in the x-, y-, and z-directions is limited to 10 m. The performance of EKF-KRMEE and the existing state estimation filters is quantified in terms of the RMSE defined as (5.25). As shown in Fig. 5.14, the estimated trajectory obtained by the MEE-based EKF, compared to the MCC-based EKF and EKF, is closer to the ground truth (GT). The observation made in Fig. 5.14 is a result of the suitability of the MEE criterion over MCC and MMSE. Furthermore, in the literature, the MCC is also claimed to tackle non-Gaussianity, and the same is inferred from Fig. 5.14, where the MCC-based EKF yields a trajectory closer to GT compared to the EKF. However, since non-Gaussianity is multimodal, the MEE-based EKF works better than the MCC-based EKF. Furthermore, in Fig. 5.15, the superiority of EKF based on MEE over EKF and EKF based on MCC is shown in terms of RMSEs of different parameters, normalized by their respective unit.

It is already shown in Figs. 5.14 and 5.15 that MEE-based EKF outperforms MCC-based EKF and EKF. Subsequently, the performance of EKF-KRMEE is compared to MEE-based EKF. In the simulations, for MEE-based EKF $\sigma = 0.95$ and for KRMEE

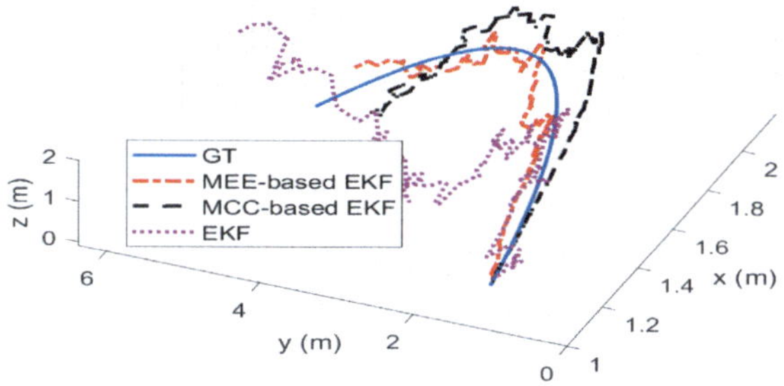

Fig. 5.14 Evolution of GT and estimated trajectories in 3D space with EKF, MCC-based EKF, and MEE-based EKF under non-Gaussian conditions

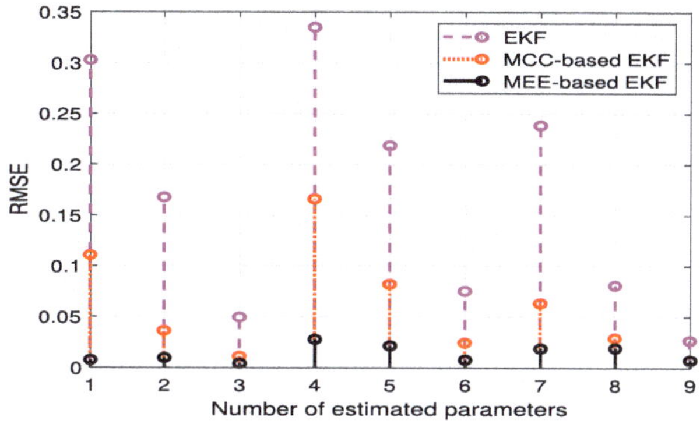

Fig. 5.15 RMSE in the estimation of positions, velocities, and accelerations using EKF, MCC-based EKF, and MEE-based EKF under non-Gaussian conditions

$\sigma_1 = 5$, also, β and λ are 1 and 0.004, respectively. Furthermore, it is depicted in Fig. 5.16 that EKF-KRMEE produces the estimated trajectory close to GT. Also, from Fig. 5.16, it can be inferred that due to estimating $\mathbf{h}(\cdot)$ and using MEE, EKF-KRMEE performs better than EKF based on MEE. Lastly, Fig. 5.17 validates the improved performance of EKF-KRMEE over MEE-based EKF in terms of the RMSE of all estimated parameters.

5.6 Conclusion

In the context of ISAC, accurate state estimation is critical for ensuring reliable performance in environments with nonlinear measurements and dynamic system behaviors. The ability to seamlessly integrate sensing and communication functionalities makes ISAC highly efficient, but it also introduces complexities in handling nonlinearities,

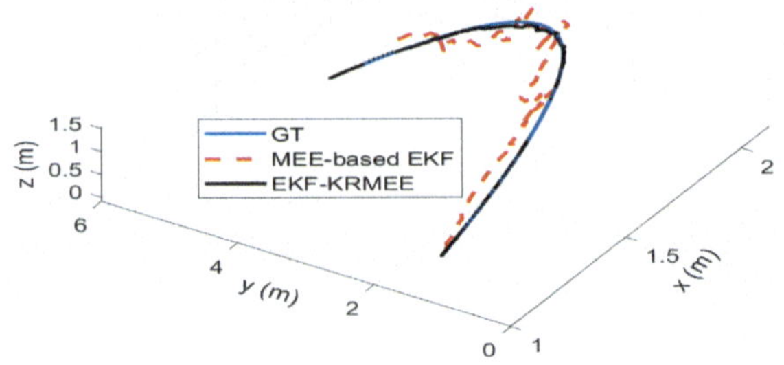

Fig. 5.16 Evolution of GT and estimated trajectories in 3D space with MEE-based EKF and EKF-KRMEE under non-Gaussian conditions

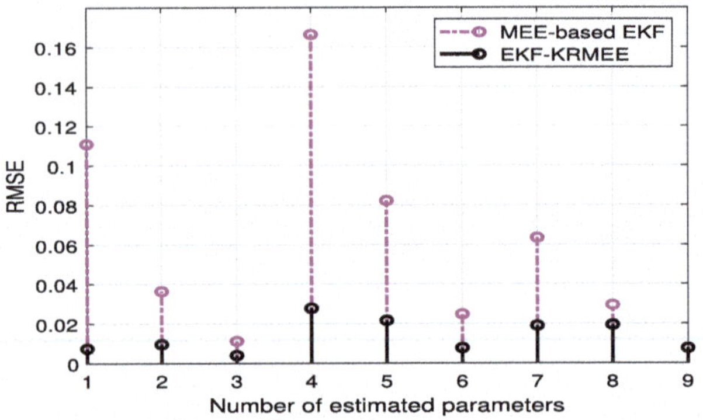

Fig. 5.17 RMSE in the estimation of positions, velocities, and accelerations using MEE-based EKF, and EKF-KRMEE under non-Gaussian conditions

non-Gaussian noise, and real-time estimation requirements. This chapter addresses these challenges by introducing advanced adaptive filtering techniques to improve state estimation accuracy in ISAC applications, validated through simulations and comparison with traditional methods.

In conventional ISAC systems, EKF and UKF are widely used to estimate states in nonlinear measurements. However, their reliance on approximating nonlinear measurement functions limits their estimation accuracy, particularly in complex environments. To overcome this limitation, the RKHS-based KRLS algorithm is applied to replace the approximation of the measurement function, resulting in the EKF-RKHS filter. This enhanced filter is particularly suited for ISAC scenarios where targets' random motion combined with spherical coordinate measurements adds complexity in state estimation. Through the use of three motion models, EKF-RKHS is shown to outperform traditional EKF, particularly in dynamic scenarios such as constant acceleration models.

In addressing dynamic parameter estimation under nonlinearity and non-Gaussianity, the EKF-MCC filter is introduced as an alternative to traditional EKF within ISAC frameworks. Unlike MMSE, MCC uses higher-order error characteristics, leading to improved estimation accuracy in noisy environments. To further enhance the capacity of EKF-MCC in ISAC, it is combined with the RKHS-based KRMC algorithm, which estimates the nonlinear measurement function directly. The resulting EKF-MCC-RKHS filter significantly improves estimation accuracy compared to EKF-MCC and conventional EKF, making it a more reliable choice for real-time ISAC applications involving dynamic target estimation.

Finally, to tackle nonlinearity in the non-Gaussian extension of EKF, the RKHS-based adaptive KRMEE algorithm is introduced for ISAC-based systems. This algorithm adapts the unknown nonlinearity using MEE at each frame and converges to an efficient estimate. The EKF-KRMEE filter, which integrates MEE-based EKF with KRMEE, demonstrates superior performance, particularly when dealing with multimodal non-Gaussian noise common in ISAC environments. This comprehensive approach provides a robust solution for state estimation in ISAC, effectively handling system nonlinearity and perturbation distributions, while ensuring high performance in real-time applications.

References

1. Bar-Shalom, Y., Fortmann, T.E., Cable, P.G.: Tracking and data association (1990)
2. Blackman, S.S.: Multiple-target Tracking with Radar Applications. Dedham (1986)
3. Dinuzzo, F., Schölkopf, B.: The representer theorem for Hilbert spaces: a necessary and sufficient condition. arXiv preprint arXiv:1205.1928 (2012)
4. Engel, Y., Mannor, S., Meir, R.: The kernel recursive least-squares algorithm. IEEE Trans. Signal Process. **52**(8), 2275–2285 (2004). https://doi.org/10.1109/TSP.2004.830985
5. Leung, H., Hu, Z., Blanchette, M.: Evaluation of multiple radar target trackers in stressful environments. IEEE Trans. Aerosp. Electron. Syst. **35**(2), 663–674 (1999). https://doi.org/10.1109/7.766946
6. Liu, W., Pokharel, P.P., Principe, J.C.: The kernel least-mean-square algorithm. IEEE Trans. Signal Process. **56**(2), 543–554 (2008)
7. Liu, W., Principe, J.C., Haykin, S.: Kernel Adaptive Filtering: A Comprehensive Introduction, vol. 57. Wiley (2011)
8. Liu, X., Qu, H., Zhao, J., Chen, B.: Extended Kalman filter under maximum correntropy criterion. In: 2016 International Joint Conference on Neural Networks (IJCNN), pp. 1733–1737 (2016). https://doi.org/10.1109/IJCNN.2016.7727408
9. Richards, M.A.: Fundamentals of Radar Signal Processing. Tata McGraw-Hill Education (2005)
10. Singh, U.K., Mitra, R., Bhatia, V., Mishra, A.K.: Kernel LMS-based estimation techniques for radar systems. IEEE Trans. Aerosp. and Electron. Sys. **55**(5), 2501–2515 (2019). https://doi.org/10.1109/TAES.2019.2891148
11. Singh, U.K., Mitra, R., Bhatia, V., Mishra, A.K.: Range and velocity estimation using kernel maximum correntropy based nonlinear estimators in non-Gaussian clutter. IEEE Trans. Aerosp. Electron. Syst. **56**(3), 1992–2004 (2020)
12. Singh, U.K., Alaee-Kerahroodi, M., Bhavani Shankar, M.R.: Implementation and validation of multiple target tracking technique for indoor applications. In: 2022 IEEE International Conference on Advanced Networks and Telecommunications Systems (ANTS), pp. 219–224 (2022). https://doi.org/10.1109/ANTS56424.2022.10227719

13. Singh, U.K., Alaee-Kerahroodi, M., Bhavani Shankar, M.R.: RKHS based dynamic state estimator for non-gaussian radar measurements. In: 2023 IEEE Radar Conference (RadarConf23), pp. 1–6 (2023). https://doi.org/10.1109/RadarConf2351548.2023.10149463
14. Singh, U.K., Alaee-Kerahroodi, M., Shankar, M.R.B., Thipparaju, R.R.: Mee-based adaptive state estimator for non-gaussian radar measurement. In: 2024 IEEE Radar Conference (RadarConf24), pp. 1–6 (2024). https://doi.org/10.1109/RadarConf2458775.2024.10548434
15. Singh, U.K., Alaee–Kerahroodi, M., Shankar, M.R.B.: RKHS based state estimator for radar sensor in indoor application. In: 2022 IEEE Radar Conference (RadarConf22), pp. 01–06 (2022). https://doi.org/10.1109/RadarConf2248738.2022.9764332
16. Wang, G., Yang, X., Wu, L., Fu, Z., Ma, X., He, Y., Peng, B.: A kernel recursive minimum error entropy adaptive filter. Signal Process. **193**, 108410 (2022). https://doi.org/10.1016/j.sigpro.2021.108410
17. Wu, Z., Shi, J., Zhang, X., Ma, W., Chen, B., Senior Member, I.: Kernel recursive maximum correntropy. Signal Process. **117**, 11–16 (2015)
18. You, H., Jianjuan, X., Xin, G.: Radar Data Processing with Applications. Wiley (2016)

Chapter 6
From Connectivity to Accessibility: Privacy-Focused Wheelchair User Detection with LTE Signal and Resolution Metrics

Sandip Jana, Amit Kumar Mishra, and Mohammed Zafar Ali Khan

Abstract To safeguard personal privacy while ensuring accessibility and safety, wheelchair detection solutions should operate without relying on intrusive sensors or video surveillance. To address this, we implement *CommSense*, a fully passive, non-invasive framework that leverages Channel State Information (CSI) from existing LTE OFDM broadcasts, eliminating any need for dedicated sensors or video monitoring. CommSense applies Principal Component Analysis (PCA) to reduce data dimensionality, then uses a Support Vector Machine (SVM) to distinguish wheelchair-induced channel perturbations from ambient multipath, delivering accurate detection without compromising personal privacy. In six real-world settings (indoor courtyards, entryways, stairwells, outdoor areas), CommSense achieves detection accuracies of 92.8–97.5% with AUC > 0.96. In this work, we extend our existing study by introducing three key enhancements: (1) comprehensive statistical characterization of the original CSI datasets, (2) controlled range experiments from 1 to 10 m that maintain detection accuracies above 88%, and (3) a novel event-centric resolution metric obtained by interpolating false-positive rate curves at a 5% threshold. Our analysis shows that the minimum inter-wheelchair separation needed for reliable discrimination grows from approximately 1.2 m at a 1 m range to over 15 m at an 8 m range. These quantitative findings not only offer concrete guidelines for tuning sensing range against spatial discrimination but also empirically demonstrate that resolution degrades as the receiver-target distance increases.

S. Jana (✉) · M. Z. A. Khan
Indian Institute of Technology, Hyderabad, India
e-mail: ee20resch11013@iith.ac.in

M. Z. A. Khan
e-mail: zafar@ee.iith.ac.in

A. K. Mishra
Faculty of Space Technologies, AGH University of Kraków, Kraków, Poland
e-mail: akmishra@ieee.org

A. K. Mishra and V. Bhatia (eds.), *Integrated Sensing and Communication for Sustainable Development Goals*, Signals and Communication Technology,
https://doi.org/10.1007/978-981-92-0209-6_6

Keywords Integrated Sensing and Communication (ISAC) · Communication-based Sensing (CommSense) · Wheelchair detection · Machine Learning (ML) · Principle Component Analysis (PCA) · Support Vector Machine (SVM)

6.1 Introduction

Wheelchairs are among the most important mobility aids for individuals with physical disabilities or age-related impairments, restoring a measure of autonomy and improving overall quality of life. Beyond simply providing mechanical support, the ability to detect when a wheelchair is present unlocks a host of smart, responsive services that can further enhance convenience, accessibility, and safety.

- **Automated access control.** In parking facilities, recognizing an approaching wheelchair can trigger barrier gates to open automatically at reserved stalls, eliminating the need for drivers to fumble for access cards or buttons.
- **Personalized wayfinding.** In large shopping malls or airports, real-time wheelchair detection can integrate with indoor navigation apps to guide users along fully accessible routes, directing them toward ramps, elevators, and wide-door entrances.
- **Smart-home adaptation.** Within the household, sensors that recognize a wheelchair can automatically adjust counter-top heights, lower kitchen cabinets, or tilt bathroom mirrors to an ergonomic level, tailoring the environment to the user's needs.
- **Proactive caregiver alerts.** In assisted-living and healthcare facilities, detecting that a resident is attempting a transfer from chair to bed (or vice versa) without assistance can trigger instant alerts to nursing staff or family members, preventing falls and ensuring timely intervention.
- **Building systems integration.** Wheelchair presence can also inform HVAC, lighting, and elevator dispatch systems, prioritizing temperature comfort, corridor illumination, and elevator car allocation to support smoother, safer traversal of public and private spaces.

By embedding automatic wheelchair detection into our everyday infrastructure, we move closer to truly inclusive design: one that not only accommodates but anticipates the needs of all users, seamlessly blending assistive technology with the built environment to foster greater independence, dignity, and well-being.

6.1.1 *Motivation*

Traditional wheelchair detection methods relying on video surveillance, infrared beams, or dedicated proximity sensors are often costly to install and maintain, raise significant privacy and data security concerns, and perform poorly under adverse lighting or weather conditions. Moreover, deploying a sensor network in every corridor, parking lot or public plaza quickly becomes logistically complex and economically prohibitive, limiting widespread adoption and leaving many critical spaces unmonitored.

By contrast, our CommSense framework exploits the ubiquity of existing LTE OFDM broadcasts to achieve *non-intrusive* wheelchair detection without any additional hardware. In essence, we "listen" to ambient downlink reference signals, already saturating urban and indoor environments and extract subtle channel state variations induced by wheelchair presence and motion. This dual-use paradigm transforms the cellular network into a passive radar, dramatically reducing capital and operational expenditures while preserving user privacy (no cameras or personal data are required). The same LTE infrastructure that carries voice and data can simultaneously support real-time assistive services across diverse venues: hospital wings, shopping malls, transit stations, and open-air parks, thereby extending the reach of accessible design with minimal incremental footprint.

6.1.2 Evolution and Applications of CommSense System

Integrated Sensing and Communication (ISAC) has recently emerged as a key enabler for beyond 5G and 6G networks, offering a unified framework that jointly optimizes data transmission and environmental perception [1]. By embedding sensing functions directly into communication waveforms, ISAC transforms wireless infrastructure into a dual-use platform that can "see" and "talk" simultaneously. While CommSense shares this overarching vision, it takes a more targeted approach: rather than redesigning physical-layer waveforms to balance throughput against radar resolution, CommSense repurposes existing LTE OFDM pilot signals under the ASIN architecture (Fig. 6.1) to deliver high-precision, event-centric sensing with minimal protocol changes.

Over the past decade our group has demonstrated the versatility of the CommSense paradigm across multiple radio standards and application scenarios. In [4], we showed that GSM broadcasts enable through-wall detection with 77.5% accuracy for human presence and 95.2% for concealed-weapon detection. Building on that foundation, [7] introduced an LTE-based passive indoor occupancy estimator, and [11] extended the approach to discriminate simultaneous multi-object scenarios. More recently, we evaluated the feasibility of applying 5G New Radio for CommSense in [13] and validated millimeter-wave (60 GHz) channel models for in situ sensing in [14]. These cumulative studies underscore CommSense's ability to leverage legacy and emerging wireless signals as low-cost, privacy-preserving sensors, paving the way for ambient-aware environments in both indoor and outdoor settings.

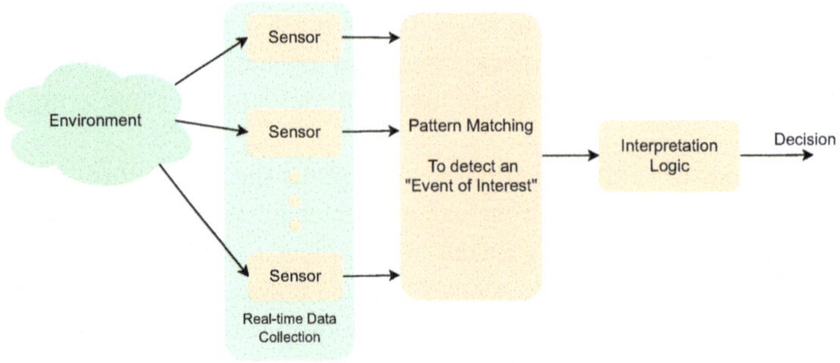

Fig. 6.1 Application-Specific Instrumentation (ASIN) is a task-driven sensor design paradigm [3, 15], inspired by echolocating species such as bats. Rather than capturing exhaustive, high-resolution data, ASIN employs machine learning to detect key environmental cues first, then performs targeted identification, filtering out noise and reducing computational load. This streamlined, bio-inspired strategy enables rapid, accurate sensing in domains like security monitoring, surveillance drones, and real-time medical diagnostics

6.1.3 Main Contributions and Structure of This Paper

This work builds upon our second experimental campaign which is an extension of the evaluations presented in I2MTC 2025 [15], to provide a comprehensive analysis of non intrusive wheelchair detection using live LTE signals. The key contributions are as follows:

- **CommSense System Design:** A passive, non-intrusive framework that leverages Channel State Information (CSI) from live LTE OFDM signals to detect wheelchair users, entirely eliminating the need for dedicated sensors or transmitters.
- **Real-World Validation:** We build upon the experiments reported in [15] to evaluate CommSense across six diverse environments—indoor courtyards, entryways, stairwells, and outdoor areas, achieving detection accuracies of 92.8–97.5% with AUC > 0.96; in this extended work, we augment those results with in-depth statistical analyses for each scenario.
- **Range Performance Characterization:** Controlled range experiments at Rx-wheelchair distances of 1–10 m with 2 m increments, showing sustained classification accuracies above 88% and quantifying performance degradation with distance.
- **Event-Centric Resolution Metric:** A Neyman–Pearson-based definition of spatial resolution, framing resolution as the minimum inter-wheelchair separation that guarantees a prescribed false-alarm rate.
- **Empirical Resolution:** A practical systematic approach for determining resolution thresholds by interpolating false-positive-rate (FPR) curves at 5%, revealing a near-linear increase in required spacing from $\approx$1.2 m at 1 m range to over 15 m at 8 m range.

- **Deployment Guidelines:** Quantitative insights into the trade-off between sensing range and spatial discrimination.

The remainder of this paper is structured as follows: Sect. 6.2 provides an in-depth analysis of the latest advancements in wheelchair detection. Section 6.3 provides a comprehensive analysis of the CommSense architecture, detailing the process from capturing LTE signals to sensing the environment. Section 6.4 presents the results from wheelchair detection in public places. Lastly, Sect. 6.5 offers a conclusion, summarizing the findings and implications of this work.

6.2 State of the Art

Several methods have been proposed to detect wheelchairs and related mobility aids using diverse sensor modalities. Huang et al. [17] develop a cascaded decision tree (CDT) approach on monocular camera images to both detect wheelchairs and infer their motion direction, achieving over 92% detection accuracy in cluttered environments.The authors in [18] present a smart wheelchair platform that embeds ballistocardiography sensors and three-axis accelerometers in the seat/backrest to unobtrusively monitor heart and respiratory rates, as well as user posture, transmitting data wirelessly via Wi-Fi and ZigBee for advanced processing. The authors in [19] propose three human–machine interfaces: an IMU-based head/neck movement system, polymeric optical fiber pressure sensors, and an SSVEP-based BCI that recognize user intent and posture to generate four-directional control commands for an electric wheelchair. An unsupervised framework by [21] applies Independent Component Analysis (ICA) coupled with a Kantorovich distance anomaly detector to recognize wheelchair users' sitting postures, reporting an F1-score of 99.41%. In the realm of 2D LiDAR data, [20] introduces Li2Former, an omni-dimensional aggregation module that fuses spatial, temporal, and batch features for enhanced people detection; similarly, [22] employ the DROW convolutional neural network with preprocessing and a voting scheme to detect wheelchairs and walkers in real time without additional sensor fusion. [23] present a depth-based pipeline combining Fast R-CNN and probabilistic estimation, reinforced by a hidden Markov model, to classify and track individuals using mobility aids in hospital settings—even under significant occlusion.

6.3 CommSense System Architecture

The CommSense framework harnesses live LTE OFDM waveforms for passive environmental sensing, without any dedicated sensing transmitters. Figure 6.2 depicts the overall signal flow, which comprises two parallel pipelines: a conventional communication chain and a sensing chain that exploits Channel State Information (CSI).

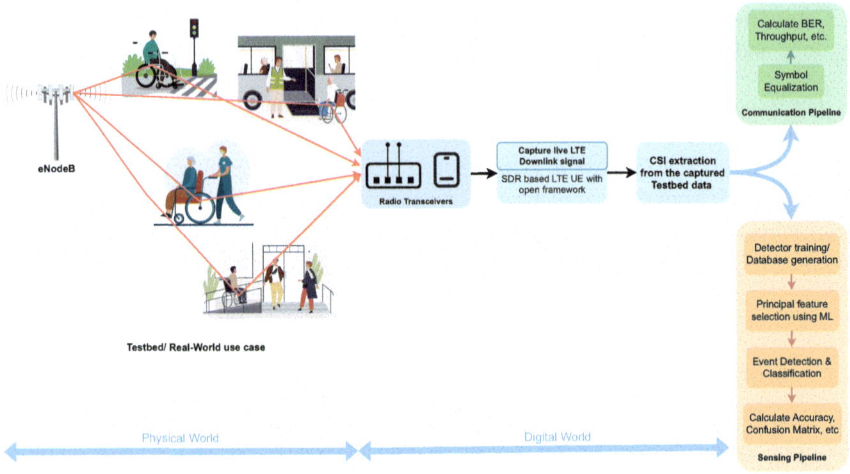

Fig. 6.2 CommSense architecture: live LTE signals are received and CSI is extracted into parallel communication and sensing pipelines, enabling simultaneous data transmission and environmental monitoring [15]

6.3.1 *Signal Acquisition*

We sample the downlink LTE signal $s(t)$ broadcast by a base station:

$$s(t) = \sum_{k=0}^{N-1} X_k \, e^{j\,2\pi k\,\Delta_f\,t}, \tag{6.1}$$

where N is the number of subcarriers, Δ_f is the subcarrier spacing (15 kHz), and X_k is the symbol on subcarrier k. The received waveform

$$r(t) = \sum_{i=1}^{L} \sum_{k=0}^{N-1} h_i \, X_k \, e^{j\,2\pi k\,\Delta_f\,(t-\tau_i)} + n(t) \tag{6.2}$$

is a superposition of L multipath reflections (each with gain h_i and delay τ_i) plus AWGN $n(t)$. From $r(t)$ we extract CSI samples over all subcarriers.

6.3.2 Communication Pipeline

Using the periodically inserted LTE pilot symbols X_p, we form the least-squares channel estimate on those subcarriers:

$$\hat{H}_p = \frac{R_p}{X_p}, \quad R_p = H_p X_p + N_p \tag{6.3}$$

We then equalize data subcarriers via $\hat{X}_k = \frac{R_k}{H_k}$ and quantify bit error rate (BER) as

$$\text{BER} = \frac{1}{n}\sum_{i=1}^{n} \mathbb{I}(x_i \neq \hat{x}_i) \tag{6.4}$$

6.3.3 Sensing Pipeline

For sensing, we leverage the full CSI matrix $\mathbf{H} = [H_1, H_2, \ldots, H_N]$ to detect environmental changes (e.g., presence of a wheelchair). The pipeline consists of

1. **Dimensionality Reduction (PCA)** Compute the covariance $\mathbf{C} = \frac{1}{M}\sum_{m=1}^{M}(\mathbf{H}_m - \mu)(\mathbf{H}_m - \mu)^T$, solve $\mathbf{C}\,\mathbf{v}_j = \lambda_j\,\mathbf{v}_j$, and project onto the top k eigenvectors:

$$\mathbf{H}_{\text{PCA}} = [\,\mathbf{v}_1, \ldots, \mathbf{v}_k\,]^T\,\mathbf{H} \tag{6.5}$$

2. **Classification (SVM)** Train a binary SVM on $\{\mathbf{H}_{\text{PCA}}, y\}$ by solving

$$\begin{aligned} &\min_{\mathbf{w},b,\xi} \ \tfrac{1}{2}\|\mathbf{w}\|^2 + C\sum_i \xi_i \\ &\text{s.t.} \quad y_i(\mathbf{w}^T\mathbf{h}_i + b) \geq 1 - \xi_i, \ \xi_i \geq 0. \end{aligned} \tag{6.6}$$

   New CSI projections $\mathbf{h}_{\text{new}}$ are labeled via $\hat{y} = \text{sign}(\mathbf{w}^T\mathbf{h}_{\text{new}} + b)$.
3. **Performance Metrics** From the confusion matrix, we compute

$$\begin{aligned} \text{Accuracy} &= \frac{\text{TP} + \text{TN}}{\text{TP} + \text{FP} + \text{TN} + \text{FN}} \\ \text{FPR} &= \frac{\text{FP}}{\text{FP} + \text{TN}}, \quad \text{FNR} = \frac{\text{FN}}{\text{TP} + \text{FN}} \end{aligned} \tag{6.7}$$

This dual-pipeline architecture enables simultaneous high-fidelity communication and passive detection of wheelchair presence.

6.3.4 *Sensing Resolution: Conventional Radar versus CommSense Redefinition*

6.3.4.1 Conventional Range Resolution

Measurement systems define *resolution* δ of an instrument as the smallest change in *feature parameter* (e.g., inter-target separation) that can be detected with a specified probability of indistinguishability from a reference standard [2]. In radar applications, this translates to the minimum detectable separation in range:

$$\Delta R = \frac{c}{2\,B\,\cos\left(\frac{\beta}{2}\right)}, \tag{6.8}$$

where

- c is the speed of light,
- B is the signal bandwidth,
- β is the bistatic angle between transmitter, target, and receiver.

Equation (6.8) applies directly when the system's goal is to *measure* the range to discrete scatterers.

6.3.4.2 Why Conventional Resolution Fails for CommSense

CommSense does not perform classical object ranging. Instead of estimating physical distances, our proposed CommSense system detects *events*: for example, "one" versus "two" wheelchairs in the scene, by monitoring changes in the estimated channel impulse response. Thus

- Our *feature parameter* (θ) is a discrete event (presence/count of objects), not an RF "distance."
- Bistatic angle β and RF bandwidth B no longer directly govern detection performance.
- We must redefine resolution in terms of the smallest change in the event of interest that remains reliably distinguishable under a fixed false-alarm constraint.

6.3.4.3 Neyman–Pearson Principle-Based Resolution

Under the Neyman Pearson (NP) framework, a CommSense detector yields a test statistic whose exceedance probability

$$P_p(\theta) = \Pr\left(\text{“event } \theta \text{ present”}\right)$$

must be maximized for a prescribed false-alarm rate

$$P_{fp} = \Pr\big(\text{declare present} \mid \text{event absent}\big).$$

We therefore define the CommSense resolution δ_{NP} as follows:

Definition (NP-Principle Resolution): The NP-principle resolution, δ_{NP}, is the largest increment in the feature parameter, $\Delta\theta$, such that the induced change in false-alarm probability, ΔP_{fp}, remains below an arbitrarily small threshold ε:

$$\delta_{\text{NP}} = \max\{\Delta\theta \mid \Delta P_{fp} \leq \varepsilon\}.$$

In practice, δ_{NP} quantifies the minimum event-to-event separation (e.g., two-wheelchair spacing) that CommSense can resolve while maintaining false alarms below ε. This event-centric definition generalizes the classical range resolution in Eq. (6.8) to any sensing modality based on hypothesis testing.

6.4 Experimental Result

6.4.1 Experiment #1: Varying the Background

6.4.1.1 Setup and Methodology

Figure 6.3a, b illustrates the deployment of our CommSense framework across two of the six distinct environments. By spanning both indoor and outdoor settings, we assess CommSense's accuracy and resilience in real-world wheelchair detection.

Figure 6.3a, b depicts two representative indoor backgrounds. In each case, an 11dBi directional antenna (center frequency 1826.5 MHz, bandwidth 20 MHz) captures the LTE downlink signal reflected from the scene. We extract Channel State Information (CSI) to quantify propagation changes induced by either a static person or a wheelchair user. From the raw CSI, we compute a set of discriminative features that highlight human presence and mobility-aid signatures. These features are then fed into our classifier to differentiate wheelchair users from other objects or activities. This end-to-end processing pipeline ensures reliable wheelchair detection across a variety of architectural layouts, supporting enhanced accessibility services, proactive healthcare monitoring, and seamless integration into smart-building infrastructures.

6.4.1.2 Average KL Divergence Across Environments

To quantify the statistical separability of the three event classes: *Ambient*, *Human* and *WhChair*; we computed the pairwise Kullback–Leibler divergence $D_{\text{KL}}(P_i \parallel P_j)$ between their empirical CSI distributions in each of the six test environments

(a) Indoor Courtyard (b) Indoor Entry 1

(c)

Fig. 6.3 **a** and **b** Depict two representative background variation scenarios from the initial experiments (Experiment #1), demonstrating CommSense's robustness to different environments. **c** Shows the empirical resolution measurement setup (Experiment #3), in which the inter-wheelchair spacing is swept at each fixed receiver distance to empirically determine the minimum resolvable separation

(two of them are shown in Fig. 6.4), then averaged these values element-wise. Figure 6.5 presents the resulting 3×3 matrix of mean divergences.

Key observations from Fig. 6.5:

- The largest average divergence is $D_{\mathrm{KL}}(\text{Ambient} \parallel \text{WhChair}) = 2.22$, indicating that the wheelchair presence distribution departs most strongly from the ambient baseline.

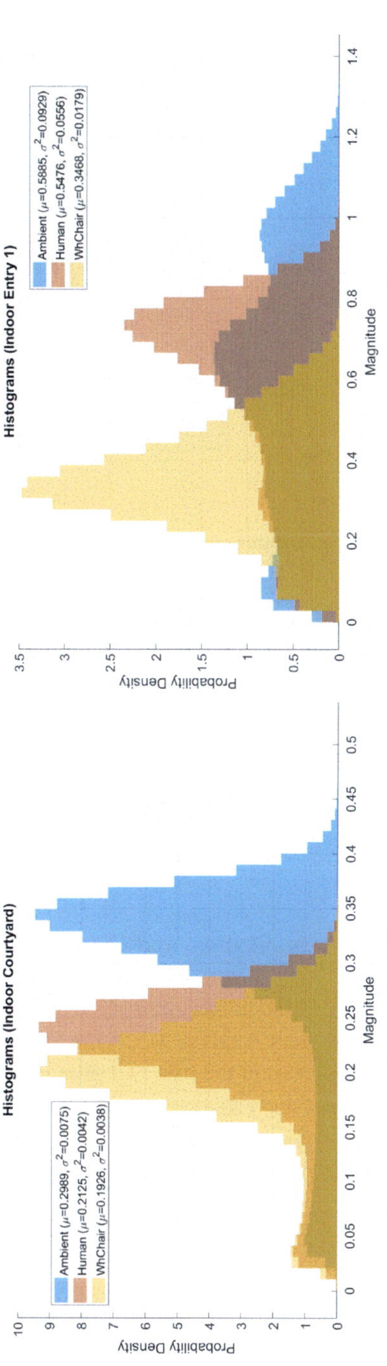

(a) Indoor Courtyard

(b) Indoor Entry 1

Fig. 6.4 Distribution of the CSI for two of the backgrounds from Experiment #1. For each of the cases we have calculated the empirical mean and variance

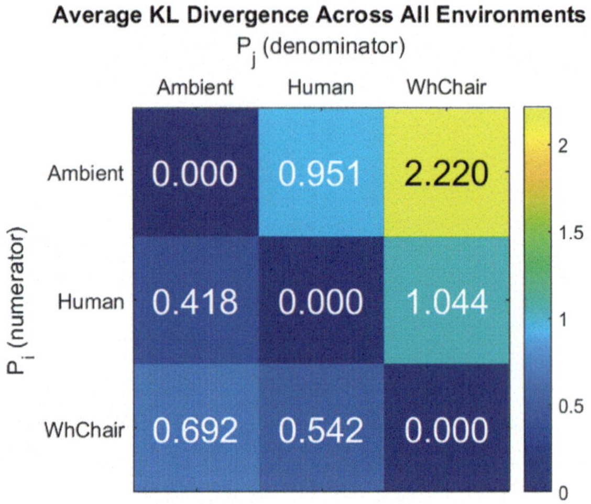

Fig. 6.5 Average KL divergence $D_{KL}(P_i \parallel P_j)$ across all environments. Rows index the "true" distribution P_i, columns index the "approximation" P_j

- By contrast, $D_{KL}(\text{WhChair} \parallel \text{Ambient}) \approx 0.69$, reflecting the inherent asymmetry of KL divergence.
- Human versus ambient divergence is moderate ($D_{KL}(\text{Ambient} \parallel \text{Human}) \approx 0.95$), while distinguishing human presence from wheelchair yields $D_{KL}(\text{Human} \parallel \text{WhChair}) \approx 1.04$ and its reverse $D_{KL}(\text{WhChair} \parallel \text{Human}) \approx 0.54$.
- All self-divergences sit at zero along the diagonal, as expected.

These averaged divergence values quantify the pairwise "distance" between classes in statistical terms. The relative magnitudes align with our classification accuracies: ambient versus wheelchair is easiest to separate, while human versus wheelchair having smaller but still significant divergence presents greater overlap.

6.4.1.3 Feature Extraction, Visualization, and Classification

To manage the high-dimensional CSI data acquired across diverse test environments, we first applied Principal Component Analysis (PCA) to identify the principal axes of variance. Projecting the samples onto the first two principal components yields a clear 2D visualization (Fig. 6.6), where distinct clusters emerge according to environmental and wheelchair-induced channel variations. For automated classification, we then selected the top four principal components, capturing the majority of the data's variance as input features to a Support Vector Machine (SVM). The SVM decision results, shown in Fig. 6.8, effectively separate wheelchair user signatures from non-user cases.

This combined PCA+SVM approach delivers robust performance across all six settings, achieving the following detection accuracies:

- Indoor Courtyard: 92.8%

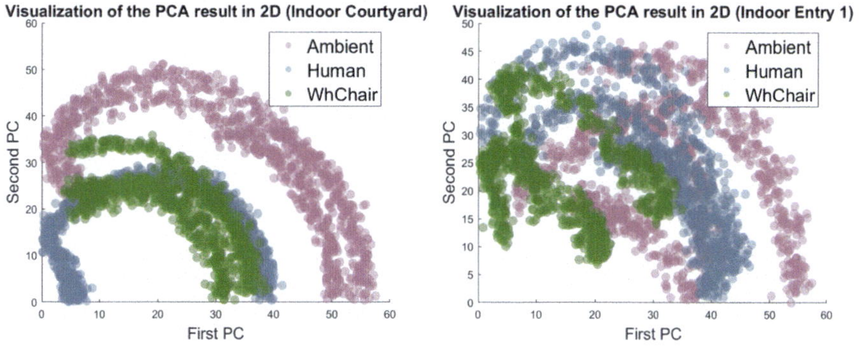

(a) Indoor Courtyard (b) Indoor Entry 1

Fig. 6.6 From two different environments the data is projected using PCA onto the two dominant eigenvectors, illustrating the variations and patterns unique to each environment

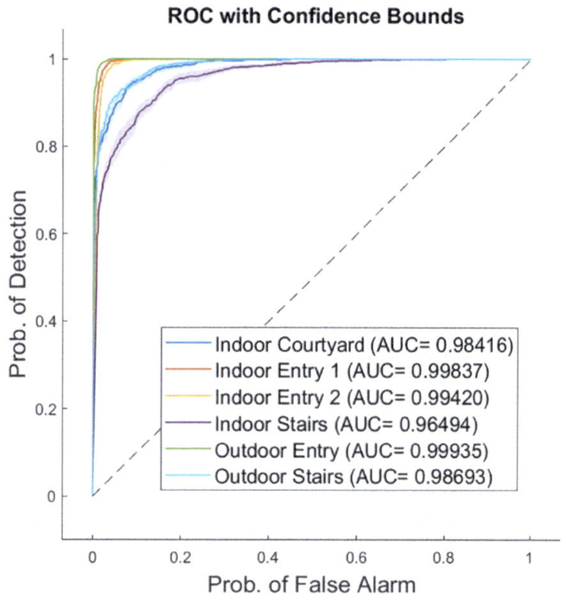

Fig. 6.7 ROC curves for all scenarios (TPR versus FPR by varying SVM threshold), with AUC > 0.96 in every case, demonstrating excellent performance

- Indoor Entry 1: 94.0%
- Indoor Entry 2: 92.8%
- Indoor Stairs: 87.1%
- Outdoor Entry: 97.5%
- Outdoor Stairs: 93.0%

These results confirm that leveraging PCA for feature extraction, followed by SVM classification, ensures high-fidelity wheelchair detection under varying multipath, clutter, and background dynamics.

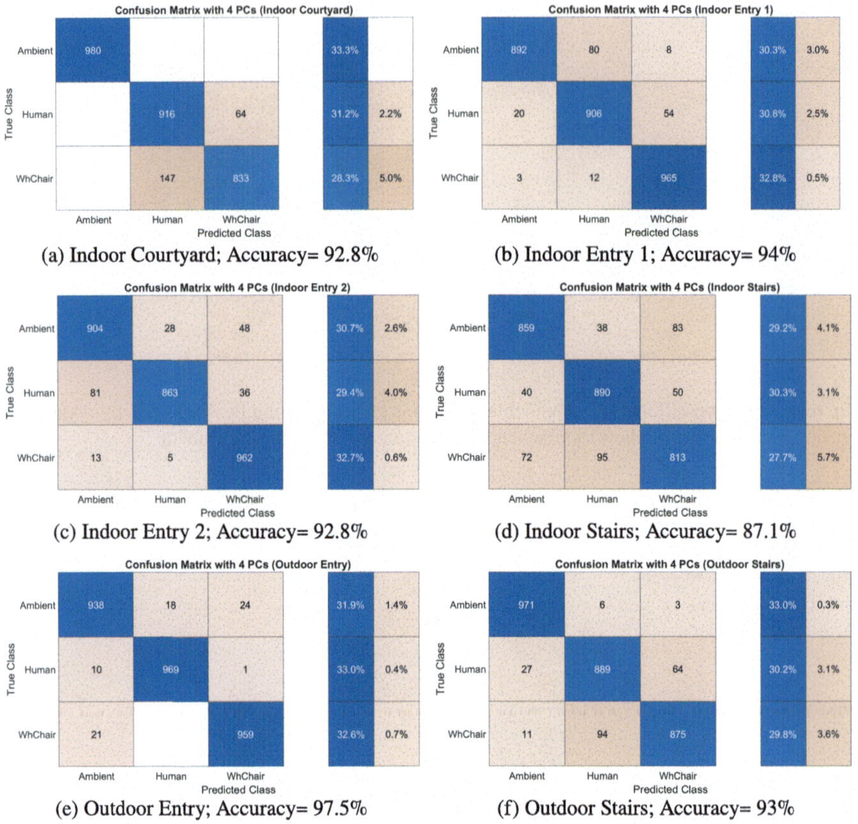

Fig. 6.8 Classification of high-dimensional estimated channel data using PCA and SVM. Data is projected onto the four dominant principal components, followed by SVM classification to distinguish between human or wheelchair presence and absence in various environments [15]

6.4.1.4 Receiver Operating Characteristics (ROC)

The receiver operating characteristic (ROC) curve serves as a fundamental metric for assessing the quality of classifiers, enabling a visual and quantitative comparison of trade-offs between true-positive rate (TPR) and false-positive rate (FPR). In our study, we swept the decision threshold of the SVM classifier for each experimental scenario to generate ROC plots, charting TPR against FPR. The resulting area under the curve (AUC) condenses overall discriminative power into a single scalar. Higher AUC values reflecting superior separation between wheelchair user and non-user classes. As illustrated in Fig. 6.7, our system achieved AUCs in the range 0.96–0.99, underscoring its excellent classification fidelity and robustness across diverse environments.

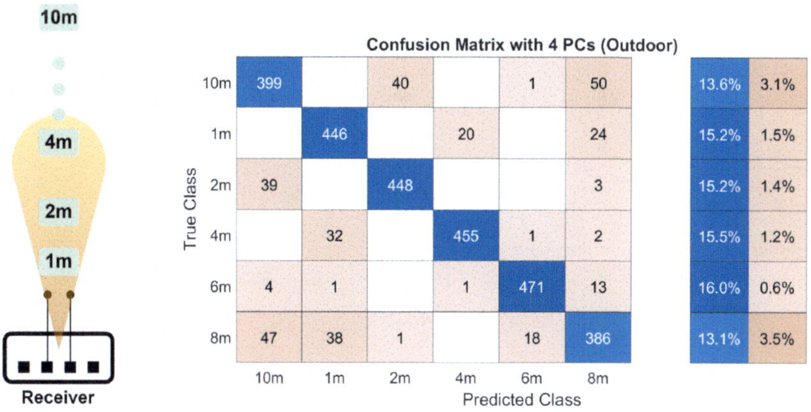

(a) Set-up: A wheelchair at increasing distances along the boresight of a fixed receiver antenna

(b) Confusion matrix (PCA–SVM with 4 PCs) across all six distances

Fig. 6.9 **a** Geometry of the measurement: the Rx-Wheelchair distance is swept from 1 to 10 m. **b** Classification results yield an overall accuracy of 88.7%, with low false-positive and false-negative rates at each range

6.4.2 Experiment #2: Performance with Varying Distance

To validate our proposed *non-intrusive* wheelchair detection approach over practical deployment ranges, we designed a second outdoor experiment in which a single wheelchair reflector was placed at six discrete radial distances from the receiver antenna: 1–10 m with 2 m increments. In each run, the wheelchair was placed within the boresight of the fixed LTE antenna beam (Fig. 6.9a), ensuring consistent signal illumination.

Figure 6.9b shows the resulting 6×6 confusion matrix when we treat each distance as a separate class and apply PCA (four principal components) followed by an SVM classifier. The overall classification accuracy is 88.7%, with true-positive rates (diagonal entries) exceeding 80% in every distance bin. Misclassifications occur most often between neighboring distances (e.g., 10 m versus 8 m), but remain below 20% for any off-diagonal entry. False-positive and false-negative rates stay under 5% for the majority of classes, indicating that the system can not only detect a wheelchair's presence but also infer its approximate range within a few meters.

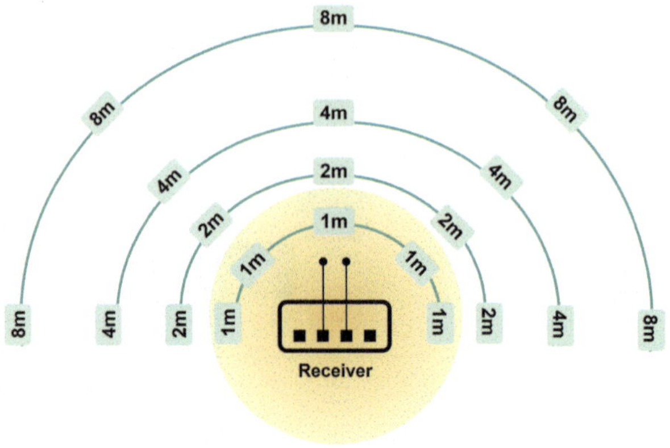

Fig. 6.10 Measurement geometry: inter-wheelchair spacing swept at each UE distance

6.4.3 Experiment #3: Resolution Performance

To quantify the minimum spatial separation at which two adjacent wheelchairs can be reliably distinguished by CommSense, we designed a dedicated systematic resolution experiment that goes beyond mere presence detection. In this phase, the task becomes a binary decision: "one" versus "two" wheelchairs; rather than "wheelchair present" versus "absent."

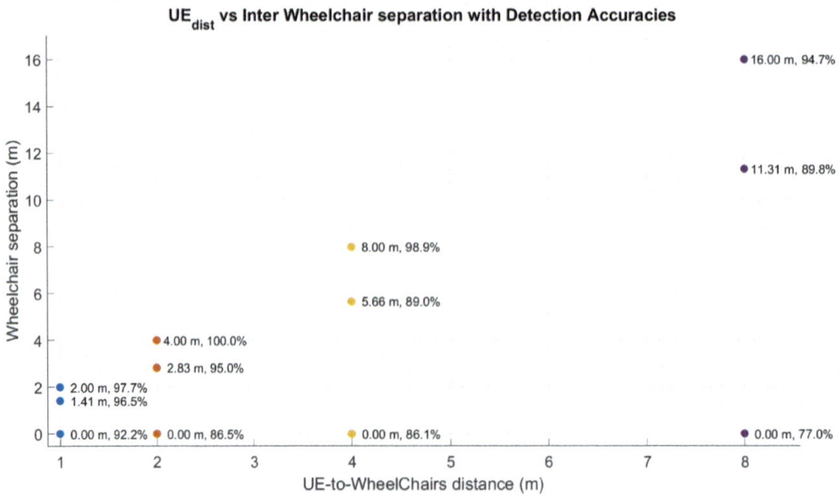

Fig. 6.11 Classification accuracy as a function of UE-to-wheelchair distance and inter-wheelchair separation. Each colored series corresponds to a fixed receiver range (1 m, 2 m, 4 m, 8 m), and each marker is annotated with its exact spacing and measured accuracy

6.4.3.1 Experimental Setup and Measurement Geometry

Figures 6.3c and 6.10 show the geometry: two wheelchairs lie at radial distances of 1, 2, 4, and 8 m. For each distance, we varied the angular offset ∈ {0°, 45°, 90°} and stepped the inter-wheelchair spacing from 0 m up to the point where classification became perfect (Fig. 6.12).

6.4.3.2 Detection Accuracy versus Separation

Figure 6.11 plots detection accuracy as a function of inter-wheelchair spacing for each receiver distance. At zero spacing, accuracies start at 92.2% (1 m), 86.5% (2 m), 86.1% (4 m), and 77.0% (8 m). As spacing increases to 1–2 m, accuracies exceed

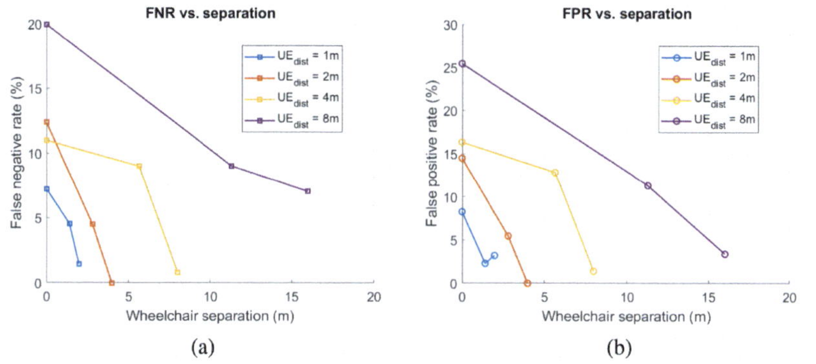

Fig. 6.12 FPR and FNR as functions of inter-wheelchair spacing, for Rx-wheelchair distances of 1, 2, 4, and 8 m

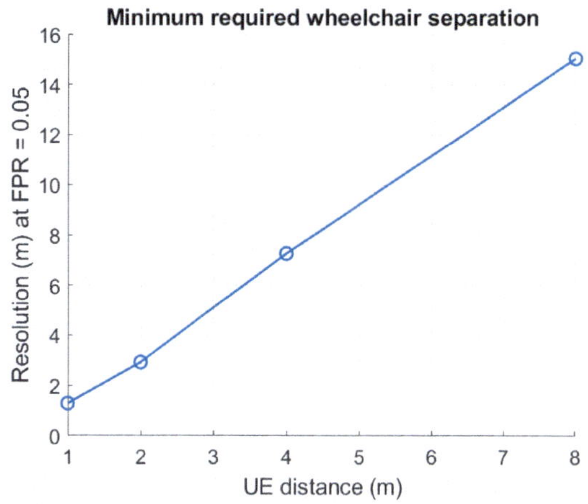

Fig. 6.13 Minimum required wheelchair separation (resolution) at an FPR of 5%, as a function of UE-wheelchair distance

95% even at longer ranges, and by 4 m separation the classifier achieves near-perfect performance (up to 100% at 2 m range and 98.9% at 4 m range).

6.4.3.3 Resolution at 5% FPR

We define the system's *resolution* as the minimum inter-wheelchair separation required to constrain FPR to 5%. By linearly interpolating each FPR curve to the 5% point, we obtained the thresholds plotted in Fig. 6.13. The resulting resolution grows approximately linearly with range (i.e., Rx-Wheelchair distance):

$$\text{Resolution} \approx \begin{cases} 1.2\,\text{m}, & d = 1\,\text{m} \\ 2.9\,\text{m}, & d = 2\,\text{m} \\ 7.2\,\text{m}, & d = 4\,\text{m} \\ 15.0\,\text{m}, & d = 8\,\text{m} \end{cases}$$

These results quantify the CommSense system's discrimination capability and inform the design trade-off between sensing range and spatial resolution in non-intrusive wheelchair detection.

6.5 Conclusion

The key contributions of this study to non-intrusive wheelchair user detection are as follows:

- Validation in six real-world scenarios using PCA+SVM, achieving 92.8–97.5% accuracy and AUC > 0.96.
- Statistical characterization of CSI via empirical PDFs and KL divergences.
- Range experiments (1–10 m) sustaining $>$88% detection accuracy.

- Empirical proof of near-linear resolution degradation with increasing range.
- Practical guidelines for balancing sensing range and spatial discrimination.

We validated CommSense in a variety of realistic scenarios, including indoor courtyards, entryways, stairwells, and outdoors. Across these diverse settings, we have done statistical analysis of the data, then we used PCA+SVM classifier that consistently achieved detection accuracies between 92.8% and 97.5%, AUC values exceeding 0.96 in ROC analysis. These results confirm the robustness and generalizability of our non-intrusive methodology, even in the presence of multipath, clutter, and changing background activity.

To characterize the system's spatial discrimination limits, we conducted controlled range experiments with the receiver positioned at distances from 1 m up to 10 m from the wheelchairs. Detection accuracy remained above 88 % in this range-based experiment, demonstrating sustained performance over practical deployment distances. By interpolating the false-positive-rate (FPR) curves at a fixed 5 % threshold, we derived the minimum inter-wheelchair separation necessary for reliable two-object resolution. This "resolution" metric increases approximately linearly with range; from roughly 1.2 m at a 1 m distance to over 15 m at an 8 m distance, quantifying the fundamental trade-off between sensing range and spatial discrimination in passive LTE-based systems.

These findings not only underscore CommSense's capability for high-fidelity wheelchair detection across varied environments, but also provide concrete guidelines for system designers: by selecting appropriate receiver-target distances and accounting for the range-resolution relationship, deployments can be tailored to meet specific spatial resolution requirements. Looking ahead, we plan to enhance CommSense through sensor fusion with additional wireless bands, explore cooperative multi-receiver configurations, and extend the framework to new passive-sensing applications.

In future work, we will explore multi-band and multi-antenna extensions of CommSense to boost spatial resolution and coverage, investigate cooperative, distributed receiver arrays for seamless area monitoring, and develop lightweight, on-edge inference algorithms to support real-time deployment.

Acknowledgements The authors gratefully acknowledge the invaluable support and insights provided by our lab mates, Vikram and Aditya, during the experimentation. We also extend our sincere thanks to Dr. Sonia and the IITH hospital staff for their cooperation.

References

1. Cui, Y., Liu, F., Masouros, C., Xu, J., Han, T.X., Eldar, Y.C.: Integrated sensing and communications: background and applications. In: Liu, F., Masouros, C., Eldar, Y.C. (eds.) Integrated Sensing and Communications. Springer, Singapore (2023)
2. Richards, M.A.: Fundamentals of Radar Signal Processing, vol. 1. Mcgraw-hill, New York (2005)

The manufacturer's authorised representative in the EU is Springer Nature Customer Service Centre GmbH, Europaplatz 3, 69115 Heidelberg, Germany. If you have any concerns regarding our products, please contact ProductSafety@springernature.com

Printed and bound by CPI Group (UK) Ltd, Croydon, CR0 4YY
07/10/2026
02229366-0002